Enhancing the Retention of Army Noncommissioned Officers

Herbert J. Shukiar

John D. Winkler

John E. Peters

Prepared for the United States Army

RAND

Arroyo Center

Approved for public release, distribution unlimited

For more information on the RAND Arroyo Center, contact the Director of Operations, (310) 393-0411, extension 6500, or visit the Arroyo Center's Web site at http://www.rand.org/organization/ard/

PREFACE

In April 1997, at the request of the Sergeant Major of the Army, RAND and the United States Army Sergeants Major Academy conducted a workshop that explored the fundamental assumptions underpinning the Army's noncommissioned officer (NCO) leader development process. This workshop's objective was to evaluate the resilience of those assumptions as the Army moves into the 21st century. Will, for example, the Army's current leader development mechanisms give NCO leaders the skills they need to address 21st-century challenges? Will the mechanisms provide future NCOs with the requisite skills?

A set of recommendations emerged from that workshop as well as an agenda for additional research. Two research areas emerged. The first addresses how the Army may provide more balance among the three pillars of NCO education: institutional, operational assignment, and self-development. The second research area focuses on NCO retention and personnel policy issues. This report presents our findings on the retention/personnel policy research area. The proceedings of the original workshop were presented in the RAND document entitled *Future Leader Development of Army Noncommissioned Officers: Workshop Results,* CF-138-A, 1998.

The research was sponsored by the U.S. Army's Deputy Chief of Staff for Personnel and the Command Sergeant Major of the U.S. Army Training and Doctrine Command and was conducted in the Manpower and Training Program of the RAND Arroyo Center. The Arroyo Center is a federally funded research and development center sponsored by the United States Army.

CONTENTS

FIGURES

TABLES

SUMMARY

RAND and the United States Army Sergeants Major Academy conducted a noncommissioned officer (NCO) leader development workshop in April 1997. The workshop's objective was to identify the assumptions that underpin the Army's current NCO leader development mechanisms and to evaluate the robustness of those assumptions as the Army moves into the 21st century. Sixty Army NCOs attended, mostly Command Sergeants Major and Sergeants Major, but with several Master Sergeants and Sergeants First Class also in attendance. Participants were drawn from all levels of the Army command structure, including the Army National Guard and the United States Army Reserve.

The workshop developed a set of recommendations and two areas that warranted further research. The first research area was concerned with providing more balance among the three NCO education pillars: institutional, operational assignment, and self-development. The second research area was concerned with NCO retention and other personnel policy practices. This report presents findings from the second research area.

MOTIVATION FOR RETENTION RESEARCH

Workshop participants expressed several concerns related to NCO retention. First, they noted that NCOs in hard-to-retain military occupational specialties (MOSs) are forced to leave the Army when they reach their retention control points (RCP), the year of service where NCOs must leave if they have not been promoted to the next-higher grade. The participants asked why these NCOs are being forced to leave when their skills are still of value to the Army.

Workshop participants also noted that NCOs often arrive at a new assignment without any training in the skills they need to perform it. These NCOs must develop those skills on the job. Further, NCOs who have served in such assignments and have the on-the-job skills must relearn those skills in a formal setting when they attend the next NCO education course.

Additionally, our meetings with senior NCOs from the Army's Enlisted Personnel Management Directorate (EPMD) elicited the observation that "getting an NCO to stay past his tenth year brought a large chance of that NCO staying for twenty."

Based on these comments, the retention research has three main thrusts. First, we examine the characteristics of NCOs and the TOE positions to which they are assigned.[1] Some NCOs are assigned to positions requiring a higher grade, some are assigned to positions requiring a lower grade, while the majority of NCOs are assigned to positions at their grade. We explore the retention and promotion characteristics of these three groups of NCOs. Second, we examine the implications of extending selected NCO RCPs in order to retain the ones who have scarce skills. Third, we examine the implications of improving mid-career NCO retention in order to get NCOs to stay past the tenth year.

We developed a spreadsheet-based, steady-state, demand-pull inventory projection model to explore the RCP relaxation and mid-career improved retention alternatives. Both RCP relaxation and retention improvement were accomplished by adjustments to the grade/year of service loss rates that are input to the model.

---

[1]We focus on TOE units rather than total authorizations because we want to see how NCO retention alternatives affect the *warfighting* Army, and TOE units carry the main warfighting burden.

RESEARCH FINDINGS

Above-Grade Fast Trackers

The Army places some NCOs in positions that are programmed for a higher grade. These above-grade NCOs (about 10 percent of E5s and E6s in TOE units are in above-grade positions) are not permitted to attend the formal NCO education course unless they have already been selected for the higher grade. Most of the above-grade NCOs have not been selected at the time they are assigned to the position. In addition, there are NCOs in the higher grade who are assigned to positions that are programmed for a lower grade, i.e., they are serving below grade.

Why are there above-grade assignments when there are NCOs with the desired grade serving below grade? Above-grade NCOs are fast trackers, and they have been placed in these positions because they have caught the eye of their superiors. That above-grade NCOs are fast trackers is demonstrated by their promotion rates. Above-grade NCOs have much higher promotion rates than their below-grade counterparts. For example, above-grade NCOs may have promotion rates that are 20 or more times higher than their below-grade counterparts. Further, above-grade NCOs have substantially higher promotion rates than their at-grade counterparts.

The Army can take steps to help these above-grade fast trackers. First, it can provide self-development mechanisms to introduce above-grade NCOs to the skills needed for their jobs, e.g., provide distance learning mechanisms to help above-grade NCOs acquire the necessary skills. Second, for those NCOs who have mastered the needed skills on the job, the NCO education system can be adjusted to provide up-front proficiency testing and enhanced curricula for those NCOs with demonstrated proficiency in the skills—in other words, don't make them learn the skills twice. Third, the Army can relax its NCO education attendance policy to allow above-grade NCOs to attend early on in their assignment.

Above-grade NCOs share one characteristic with their at- and below-grade counterparts. They leave the Army at about the same rate. The question: How can the Army encourage these fast-tracking NCOs to remain in the Army?

Relaxing E7/E8 RCPs

How will the experience mix of the NCO corps be affected if RCPs are relaxed, either outright or on a selective basis? Given the few NCOs in the junior grades who reach their RCP years, it makes sense to relax only E7 and E8 RCPs. Allowing 50 percent of these NCOs to stay three years beyond their RCP year improves the NCO experience levels by about 2 percent (there is a 2 percent increase in the number of NCOs serving beyond ten years). Allowing 90 percent to stay three years beyond their RCP improves experience levels by 4 percent.

Coupled with this increase in NCO experience levels is a modest decline in promotion rates. Promotion rates in the junior grades decline. The promotion rate to E7 declines from 5 to 9 percent, and to E6 from 2 to 3 percent. This causes the number of senior E5s and E6s, i.e., those with more than ten years of service, to increase: 18 to 28 percent for E5s and 7 to 12 percent for E6s. These increases result because of reduced promotions to E7—the additional senior E5s and E6s are those who were not selected for promotion.

Improving Mid-Career E5 and E6 Retention

What if we improved mid-career (years seven through ten) retention by 25 percent? . . . by 50 percent? E5 and E6 are the only grades for which this makes sense. What if we targeted this retention improvement to the above-grade E5s and E6s with seven to ten years of service? NCO experience levels increase by 6 to 12 percent. Further, E6 experience levels increase by 11 to 22 percent, and E5 experience levels increase by 84 to 178 percent. Unlike the E5 and E6 increases that result from relaxing E7/E8 RCPs, these E5s and E6s are fast trackers, the ones to whom the retention-improvement strategies are targeted.

Junior NCO promotion rates drop modestly. Promotion rates to E6 drop by 2 to 5 percent, and those for E5 drop by 3 to 6 percent. However, promotion rates to E7 increase by 8 to 17 percent.

Relaxing E7/E8 RCPs Versus Improving E5/E6 Mid-Career Retention

Improving E5 and E6 mid-career retention has quantitative and qualitative advantages over E7/E8 RCP relaxation. Mid-career retention improvements affect experience levels all along the senior years of service (from 11 through 20 years and beyond), and those improvements result in the retention of fast-tracking E5s and E6s. The experience improvements that emerge from E7/E8 RCP relaxation come primarily from the increased years of service the E7s and E8s are allowed to serve, roughly years 22 through 28. There are E5 and E6 experience improvements, but these come from E5s and E6s who failed to win promotion and not from fast-tracker E5s and E6s.

Cost Concerns

Any move to improve E5/E6 mid-career retention or extend E7/E8 RCPs has cost implications that should be considered, and our research has not examined these costs. For example, selectively relaxing E7/E8 RCPs or improving E5/E6 mid-career retention inevitably lead to a more senior force, with increased compensation and retirement costs. Improving E5/E6 mid-career retention would also require some form of bonus, selectively available to above-grade NCOS in hard-to-retain and/or high-tech MOSs. These cost increases would be mitigated by reduced accession and training costs, especially when mid-career retention improvement is the objective.

CONCLUSIONS

The analysis clearly points to improving fast-tracker E5/E6 mid-career retention as having a higher impact on the force structure than selectively relaxing E7/E8 RCPs. Improving the retention behavior of mid-career fast trackers in hard-to-retain or high-tech MOSs focuses on the NCOs the Army wants to retain. The fact that their retention behavior mirrors that of their at- and below-grade colleagues should make them prime targets for retention-improvement efforts. Such efforts should include accelerated education opportunities, improved self-development venues, and financial incentives.

ACKNOWLEDGMENTS

Special thanks go to Major Doug Hersh and Lieutenant Colonel Steve Galing, both of the U.S. Army DCSPER's Military Strength Analysis and Forecasting Directorate, as well as Lieutenant Colonel Harold Hardrick and Lieutenant Colonel James Harrison, for their generous help in providing out-year MOSLS results at the enlisted aggregate and the CMF level of detail. These results permitted the development of input data for the inventory projection model used to conduct the RCP relaxation and mid-career retention analysis presented in this report.

Thanks also go to RAND colleagues Robert Kerchner and Henry (Chip) Leonard for their careful technical reviews of an earlier draft. Their efforts helped to make the document clearer and more accurate.

ABBREVIATIONS

ANCOC	Advanced Noncommissioned Officer Course
BNCOC	Basic Noncommissioned Officer Course
CMF	Career Management Field
DSL	Duty Skill Level
EMF	Enlisted Master File
EPMD	Enlisted Personnel Management Directorate
IPM	Inventory Projection Model
MOS	Military Occupational Specialty
MOSLS	MOS Level System
NCO	Noncommissioned Officer
NCOES	Noncommissioned Officer Education System
PERSCOM	Total Army Personnel Command
PLDC	Primary Leadership Development Course
PMAD	Personnel Management Authorization Document
PSL	Primary Skill Level
RCP	Retention Control Point
TTHS	Trainees, Transients, Holdees, and Students
TOE	Table of Organization and Equipment
YOS	Year of Service

Chapter One

INTRODUCTION

In April 1997, at the request of the Sergeant Major of the Army, the United States Army Sergeants Major Academy and RAND conducted a noncommissioned officer (NCO) leader development workshop at Fort Bliss, Texas. The workshop, attended by about sixty senior Army NCOs from a cross section of Active and Reserve Component organizations, had as its primary goal the identification of NCO leader development areas that need evaluation and possible revision as the Army NCO corps moves into the 21st century.[1]

RESEARCH MOTIVATION: TWO MAJOR RESEARCH AREAS

This workshop identified two major areas worthy of additional research. The first area deals with the NCO education system (NCOES) directly, and especially the interrelationships among the three NCOES pillars: the institutional or formal education pillar, the operational or on-the-job training pillar, and the self-development pillar.

The second major research area deals with personnel management and related retention issues: retention control point relaxation and mid-career retention improvement for high-tech and hard-to-retain military occupational specialties (MOSs). These issues emerged in part from the NCO leader workshop and in part from post-workshop

[1]See John D. Winkler et al., *Future Leader Development of Army Noncommissioned Officers: Workshop Results*, Santa Monica, CA: RAND, CF-138-A, 1998, for a complete treatment of the workshop and its findings.

discussions with senior NCOs in PERSCOM's Enlisted Personnel Management Directorate (EPMD). The NCO workshop attendees called for relaxation of NCO retention control points (RCPs) in hard-to-retain and high-tech MOSs. They expressed concern that forcing qualified NCOs to separate when they reach their RCPs is counter-productive, especially as the Army begins to exploit higher-technology systems and places increasing demands on hard-to-retain MOSs. Relaxing RCPs would allow the Army to continue to benefit from the experience of these senior NCOs.

NCO workshop attendees also noted that NCOs frequently show up for assignments without the formal NCOES schooling/training needed to perform those assignments. The skills these NCOs should have developed in a formal NCOES setting must be learned on the job. Further, when these NCOs ultimately attend the formal NCOES course (e.g., BNCOC, ANCOC), they are retaught the skills that they've already learned on the job.

EPMD NCOs noted that once an NCO makes it to ten years of service, there is a high likelihood that he'll stay in the Army until twenty years of service. This suggests that it may be worthwhile to try to influence the NCO's mid-career retention behavior. For example, what effect would a 25 percent improvement in mid-career retention have on the NCO corps' experience profile?

RESEARCH APPROACH

This report presents the major personnel management and retention research findings. We examine the empirical basis, if any, for the workshop attendees' concerns about RCP force-outs. What, for example, would be the effect of extending RCPs on a limited basis? How would the experience mix of the NCO corps be altered by such an extension?

We also examined the retention and promotion profiles of recent years' NCO corps in order to understand the retention and promotion characteristics of mid-career NCOs and to understand the characteristics of NCOs assigned to positions for which they have no formal schooling/training. We examined the one-year separation and promotion rates of NCOs who were assigned to TOE units in November 1996.[2] To analyze separation and promotion characteristics, we divided these NCOs into three groups: those serving above their primary skill levels, those serving at their primary skill levels, and those serving below their primary skill levels: above-, at-, and below-PSL, respectively. Those serving above-PSL held positions in November 1996 that required someone with a higher primary skill level than the position holder, i.e., someone with a higher grade. Those serving at-PSL held positions in November 1996 requiring someone with the primary skill level of the holder, i.e., with the same grade as the position holder. Those serving below-PSL held positions in November 1996 requiring someone with a lower primary skill level than the position holder, i.e., with a lower grade than the position holder. The analysis examined the one-year separation and promotion experience of each group in each NCO grade.[3]

To examine how selective relaxation of RCPs and changing mid-career retention behavior affect the NCO corps' experience profile, we developed a steady-state inventory projection model (IPM) of the enlisted force.[4] This model is populated with promotion and retention data for the aggregate enlisted force and for selected enlisted career management fields (CMFs). The data were derived from out-year projections of the MOS Level System (MOSLS), a dynamic inventory projection model of the enlisted force at the MOS level of detail.[5]

---

[2]November 1996 was chosen for two reasons. First, we wanted a point in time where we could look at least a year into the future, and the most recent EMF available at the time this research was conducted was that for November 1997. Second, we wanted a point in time that was far enough beyond the military drawdown so as to be only minimally influenced by it. Our focus on TOE units stems from our desire to examine how retention affects the operational, warfighting Army.

[3]To examine NCO one-year retention and promotion behavior, we integrated Enlisted Master File (EMF) data for each NCO in a TOE position in November 1996 with data about the position's required grade taken from the Personnel Management Authorization Document (PMAD).

[4]The model holds the inventory in each grade constant and determines the promotion flows needed to support that inventory under various retention alternatives. The model's formulation is presented in Appendix B.

[5]MOSLS has been developed and maintained by GRC International, Inc., for the U.S. Army DCSPER's Military Strength Analysis and Forecasting Directorate. The model's mathematical logic is documented in *Documentation Updates for Mathematically*

SUMMARY OF RESEARCH FINDINGS

The above-PSL NCOs had much higher one-year promotion rates than their at- and below-PSL counterparts. They are fast trackers. However, their one-year separation (from the Army) rates were about the same as those of their at- and below-PSL counterparts. The question: What can the Army do to encourage these fast trackers to stay in the Army?

For most NCO grades, but primarily for E5s and E6s, the NCOs serving above-PSL were not eligible to attend the requisite NCOES course because they had not yet been selected for promotion to the grade for which the course was tailored. These NCOs, therefore, had to learn on the job the skills needed for these above-PSL assignments.

Three measures of effectiveness were used to evaluate personnel policy changes: (1) the number of NCOs serving beyond ten years of service, (2) the average time in grade for each NCO grade, and (3) the probability of promotion into each NCO grade.[6] Although improving mid-career retention and RCP extension both lead to more NCOs serving beyond ten years and to increased average times in grade, improving mid-career retention provides more dramatic improvements than does extending RCPs. Additionally, working to retain E5s and E6s at their mid-career points to get them to "stay to 20," and specifically targeting the above-PSL E5s and E6s, is more likely to retain fast-tracker above-PSL NCOs than would be the case were E7 and E8 RCPs extended.[7]

Mid-Career Retention

E5s and E6s in TOE units who serve above-PSL have much higher one-year promotion rates than do their at-PSL counterparts. Fur-

---

Complex Programs in ELIM, MOSLS and OPALS, September 1996, written by GRCI as part of the Strength Management Systems Redesign (SMSR).

[6]The probability of promotion into grade E_n is the probability, given that an NCO was just promoted into grade E_{n-1}, that he will be promoted to grade E_n sometime during his career.

[7]Getting fast-tracking E5s and E6s to stay to 20 could also lead to increased pay and retirement costs, as well as reduced accession and training costs. This analysis has not examined those cost issues.

thermore, those serving below-PSL have much lower one-year promotion rates than do their at-PSL counterparts. Indeed, those serving above-PSL are fast trackers, in their mid-career years of service. Interestingly, when looking at one-year retention behavior, there is not much difference in one-year retention rates for those serving above-, at- and below-PSL.

This suggests that an improvement in mid-career retention (retention in the 7th through 10th years of service), aimed specifically at those E5s and E6s serving at- or above-PSL, can have an important effect on the NCO corps' experience profile.[8] For CMF 67 (aircraft maintenance), improving E5 and E6 mid-career retention by 25 percent[9] leads to an 84 percent increase in the number of E5s serving beyond ten years of service and an 11 percent increase in E6s serving beyond ten years. This equates to a 6 percent increase in CMF 67 NCOs serving beyond ten years. Improving CMF 67 mid-career retention by 50 percent leads to a 178 percent and 22 percent increase respectively in E5s and E6s serving beyond ten years of service, or a 12 percent increase in total inventory serving beyond ten years.

While the number of CMF 67 NCOs serving beyond ten years increases with an increase in mid-career retention, the average times in grade for E5s and E6s also increase. The E5 average time in grade increases by 5 and 12 percent respectively for a 25 percent or 50 percent improvement in mid-career retention. The average E6 time in grade increases by 8 and 17 percent respectively for a 25 or 50 percent improvement in mid-career retention.

However, just as average time in grade increases, E5 and E6 probabilities of promotion (and the number of promotions) to these grades

---

[8]We hold the inventory in a grade constant. Mid-career retention improvement for a grade results in a redistribution of the NCO inventory over the years of service in that grade, with fewer in the earlier years and more in the later years. We have assumed that those above-PSL E5s and E6s who were induced to remain in the force past the tenth year will have the same retention behavior during their 11th through 20th years as do other NCOs in those years of service.

[9]The IPM uses grade/year of service loss rates to move the inventory from one year of service to the next, the retention rate for a grade/year being the complement of the loss rate. To achieve a 25 percent improvement in retention, E5 and E6 loss rates in years 7–10 are multiplied by 0.75. A 50 percent retention improvement is achieved by multiplying the associated loss rate by 0.5. See Appendix C, Figure C.8.

drop from 3 to 6 percent, depending on the improvement in mid-career retention. This is because E5s and E6s are staying longer, so fewer promotions into E5 and E6 are needed. However, this approach also leads to an 8–17 percent increase in probability of promotion to E7, i.e., those E6s who stay longer have more of a chance to make E7.[10]

RCP Relaxation

Relaxing RCPs makes sense only for the grades of E7 and E8. Too few E5s and E6s stay in the Army through their RCP years, while substantial numbers of E7s and E8s are forced out at their RCP years.

We expect that not all E7s and E8s who reach their RCP years are worthy of retention beyond their RCPs. If we keep 50 percent for three years beyond their RCP year (with 10 percent attrition during each of the extension years), E5, E6, and force-wide increases in NCOs beyond ten years are 15 percent, 7 percent, and 3 percent respectively. This is coupled with a 1–3 percent increase in E5 and E6 average time in grade, a 9 percent increase in E7 average time in grade, and a 5 percent increase in E8 average time in grade. This is also accompanied by a 2–5 percent reduction in probability of promotion to E5 and E6, along with a 4–5 percent increase in the probability of promotion to E7 and E8.

ORGANIZATION OF THIS REPORT

Chapter Two shows the results of analyses examining relationships among NCO grade, assignments, promotions, and retention. Chapter Three examines in detail the effects on the force structure of improved mid-career retention and RCP extension. Chapter Four summarizes our research findings.

---

[10]E5, E6, and E7 inventories are being held constant. Thus, an improvement in E5 and E6 retention has the effect of increasing the average time in grade for E5 and E6, thereby leading to reduced promotions to E5 and E6. Since E7 retention is not being changed, the number of promotions needed to support the E7 inventory does not change, which leads to an increased probability of promotion to E7.

Chapter Two

PRIMARY-SKILL-LEVEL DATA ANALYSIS

The NCO Leader Development Workshop attendees noted that NCOs often arrived at a duty station without the formal schooling required to perform the assignment. These NCOs were required to develop the needed technical and leadership skills on the job. Further, when they subsequently attended the formal NCOES course, the curriculum included the formal schooling on those skills. Part of our research effort focused on the degree to which this situation actually occurs in the NCO corps. We also examine the differences between NCOs serving above-PSL, at-PSL, and below-PSL.

SELECT-TRAIN-PROMOTE AND ABOVE-PSL NONCOMMISSIONED OFFICERS

Several years ago, as a cost-saving method, the Army implemented the *select-train-promote* NCOES eligibility policy. This policy requires an NCO to be on a selection list for promotion to the next-higher grade before he will be allowed to attend the NCOES course associated with that grade. Thus, an E4 must have been selected for E5 before he can attend the Primary Leadership Development Course (PLDC). An E5 must be on the E6 selection list before he can attend the Basic Noncommissioned Officer Course (BNCOC). An E6 must be on the E7 selection list before he can attend the Advanced Noncommissioned Officer Course (ANCOC), and an E8 must be on the E9 selection list before he can attend the Sergeants Major Course (SMC).

Yet it is in PLDC, BNCOC, and ANCOC that NCOs are formally introduced to the skills they will need to serve in the associated grade's positions. An NCO serving above-PSL who is not on the selection list for the next grade cannot, because of *select-train-promote,* attend the formal NCOES course associated with the position he holds. Figure 2.1 illustrates this. For NCOs serving in TOE units in November 1996, this chart shows by grade the number serving above-, at-, and below-PSL as a percentage of the total number in the grade. The numbers in each category by grade are shown at the figure's bottom. The total number of NCOs illustrated in the chart is 171,438—i.e., 171,438 enlisteds (grades E4–E8) served in TOE units in November 1996.[1]

About 10 percent of E4s through E6s serve above their PSL, and about half as many (in percentage terms) E7s do. Virtually no E8s

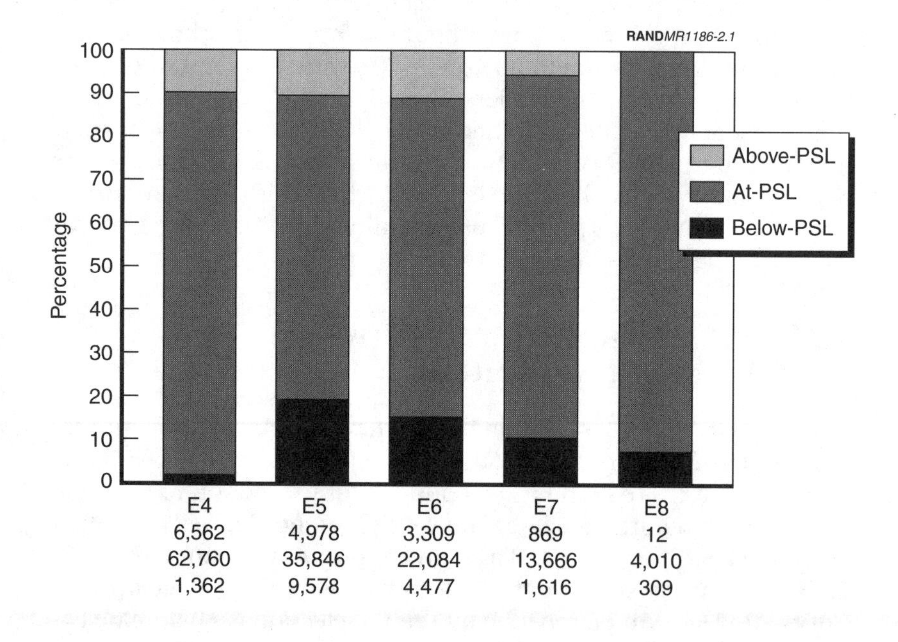

Figure 2.1—FY96 DSL/PSL Match-Up

[1]Some of the above-PSL E4s through E6s are probably on the selection list for the next grade—see Figure 2.10 and the related discussion. However, given that above-PSL E4s, E5s, and E6s have about the same one-year separation rates as their at- and below-PSL counterparts, the question remains as to how the Army might persuade them to stay.

serve above-PSL. Interestingly, along with the E4s through E7s who serve above-PSL, we see E5s through E8s serving below-PSL. For example, while 6,562 E4s serve above-PSL, 9,578 (just under 20 percent) E5s serve below-PSL.

Are Above-PSL NCOs Different from Their At- and Below-PSL Counterparts?

Why are there NCOs serving above-PSL at the same time that higher-grade NCOs are serving below? Those serving above-PSL are fast trackers, i.e., NCOs who will move up to higher grades, and those serving below-PSL are not. This is shown when we look at the one-year promotion rates for the above- and below-PSL NCOs, seen in Figures 2.2 and 2.3.[2]

E6s serving above-PSL during years 7–10 have a one-year promotion rate of about 16 percent, while their at-PSL counterparts' rate is about 5 percent, and their below-PSL counterparts have one-year promotion rates of about 1 percent. This indicates that above-PSL E6s are indeed fast trackers relative to their at- and below-PSL counterparts.

Surprisingly, the one-year separation rates for above-, at-, and below-PSL E6s in years 7–10 are the same, which is also the case for the E5s. Why, given that above-PSL E5s and E6s get promoted at much higher rates than do their at- and below-PSL counterparts, do they leave the force at about the same rate? What can the Army do to retain these fast-tracking NCOs?

Figures 2.2 and 2.3 focus on E5s and E6s with 7–10 years of service. What about *all* E5s and E6s? Do they display the same behavior, i.e., are all above-PSL NCOs fast trackers when compared with at- and below-PSL NCOs? Figures 2.4 and 2.5 show E6 and E5 one-year retention and promotion behavior over all their years of service.

These figures clearly indicate that *all* E5s and E6s who served above-PSL in November 1996 are promoted at much higher rates than their

---

[2]These figures and subsequent ones contain the expressions DSL < PSL, DSL = PSL, and DSL > PSL. DSL refers to duty skill level, and is associated with a position. An NCO serving above-PSL is in a position whose DSL is higher than the NCO's PSL. At- and below-PSL NCOs are in positions whose DSLs are at or below the NCO's PSL.

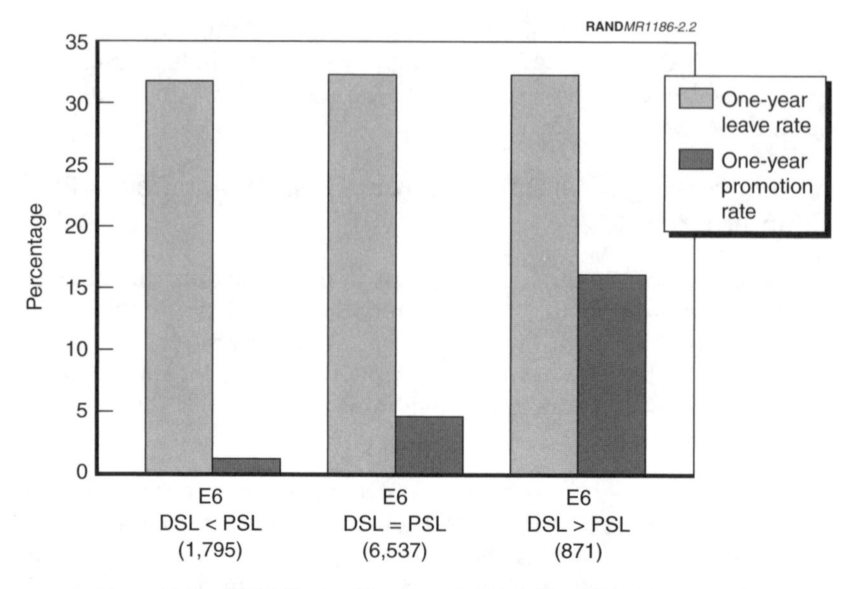

Figure 2.2—FY96 Year of Service 7–10 E6 One-Year Leave and Promotion Rates

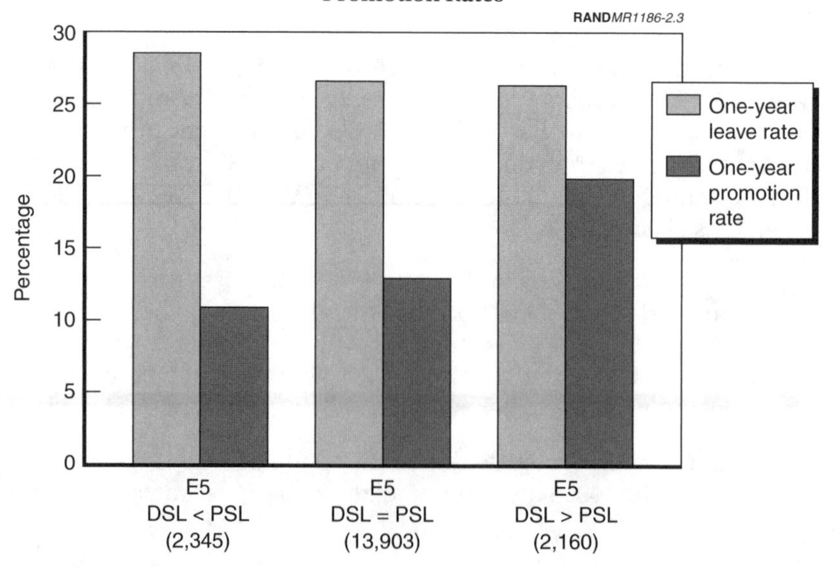

Figure 2.3—FY96 Year of Service 7–10 E5 One-Year Leave and Promotion Rates

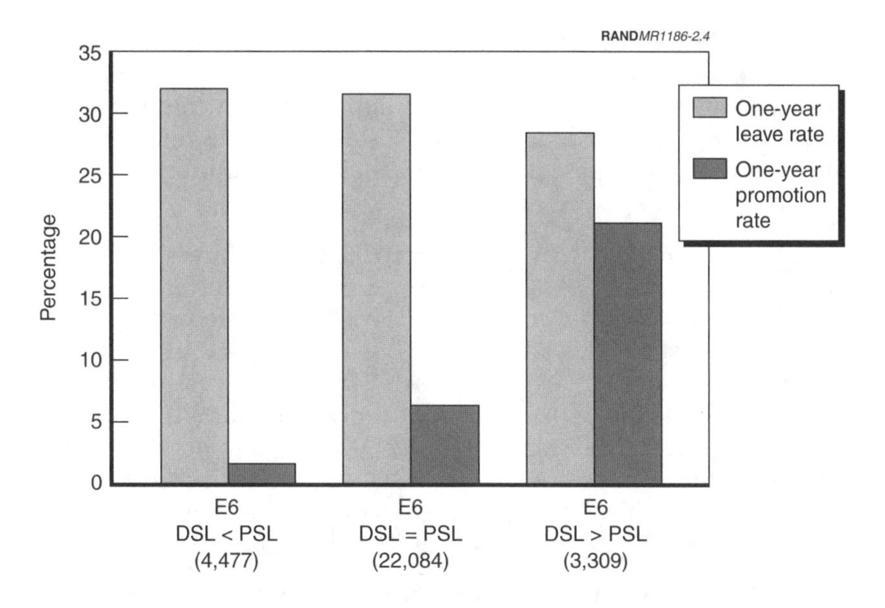

Figure 2.4—FY96 E6 One-Year Leave and Promotion Rates

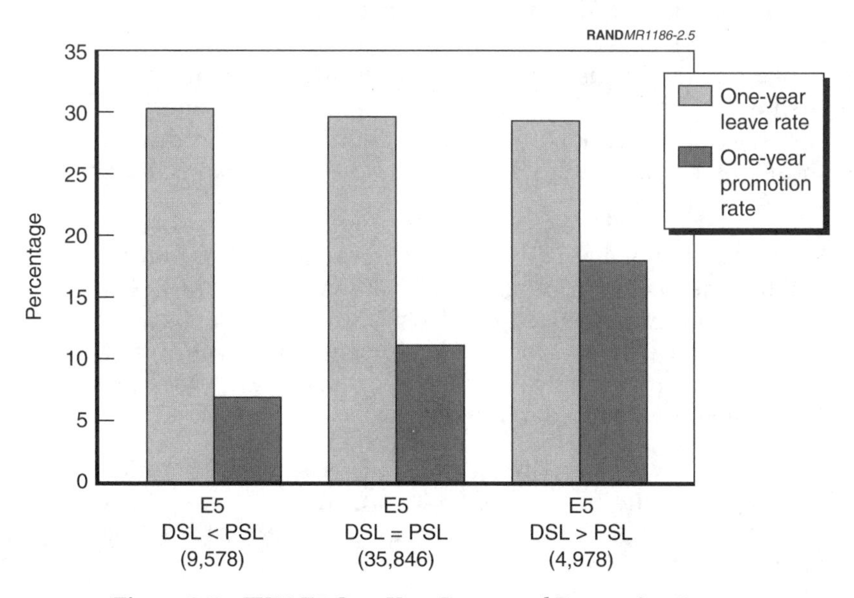

Figure 2.5—FY96 E5 One-Year Leave and Promotion Rates

at- and below-PSL counterparts. Further, the one-year separation rates are about the same, irrespective of the NCOs' serving above-, at-, or below-PSL. The numbers in parentheses at the bottom of each figure are the total number of E6s or E5s in the category, and these numbers compare exactly with those at the bottom of Figure 2.1.

Is 1996 an anomalous year? Will we see the same effects if we look at other years? Figures 2.6 through 2.9 show the same charts for November 1994 and November 1992. While not as dramatic as FY96, these charts show that above-PSL E5s and E6s are promoted at higher rates than their at- and below-PSL counterparts. Above-PSL promotions take place at two to four times the rate as those for at- and below-PSL NCOs. Above-PSL NCOs leave at a modestly lower rate than do their at- and below-PSL counterparts. It is clear from these charts that FY96 is not an anomalous year, although the FY96 promotion rate differences are larger than those for FY92 and FY94.

How do the one-year promotion and separation rates vary across CMFs? Table 2.1 shows E6 one-year promotion and retention rates (not separation rates) for those NCOs serving in TOE units in November 1996. With some interesting exceptions, the E6s serving above-PSL have substantially higher one-year promotion rates than their at- and below-PSL counterparts. Further, the one-year retention rates don't vary by much. The figure's shaded rows indicate those CMFs that have small numbers of E6s serving in TOE units.

CMF 18, Special Forces, is an interesting exception. First, there are no E6s serving below-PSL (because there are no E5 authorizations), and the one-year promotion rates for at- and above-PSL E6s are fairly close (20 and 23 percent respectively). Since this CMF is a *derivative* CMF, staffed with NCOs who have transferred from other CMFs rather than with directly accessed enlisteds, the Army can be selective about those allowed to transfer.

Some CMFs have very low above-PSL one-year promotion rates: CMF 74, Automatic Data Processing; CMF 95, Military Police; and CMF 98, Signals Intelligence/Electronic Warfare. Each of these CMFs has a one-year promotion rate below 10 percent.

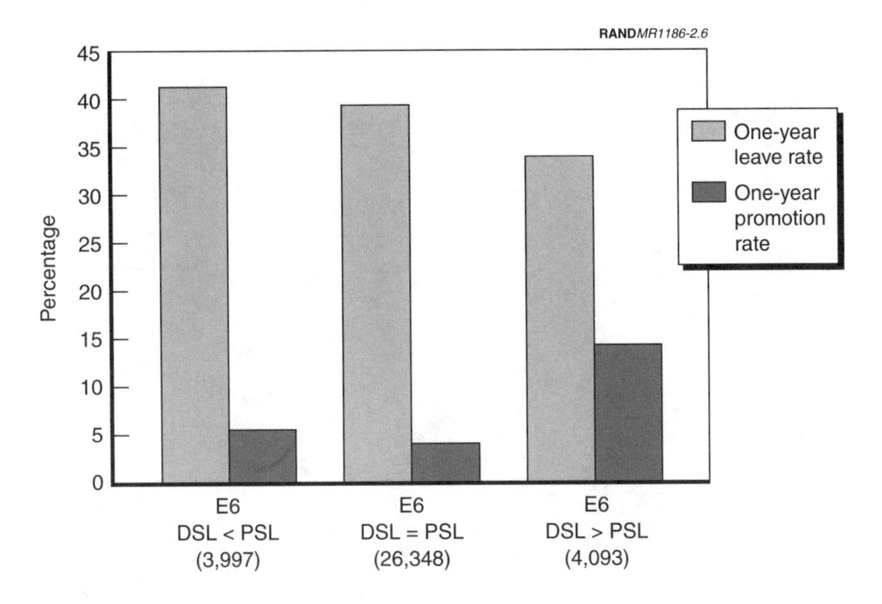

Figure 2.6—FY94 E6 One-Year Leave and Promotion Rates

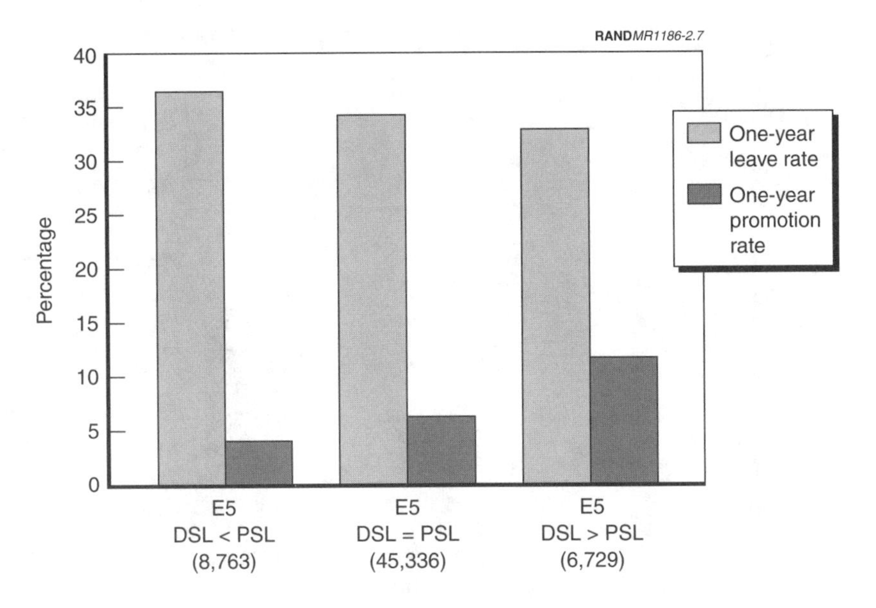

Figure 2.7—FY94 E5 One-Year Leave and Promotion Rates

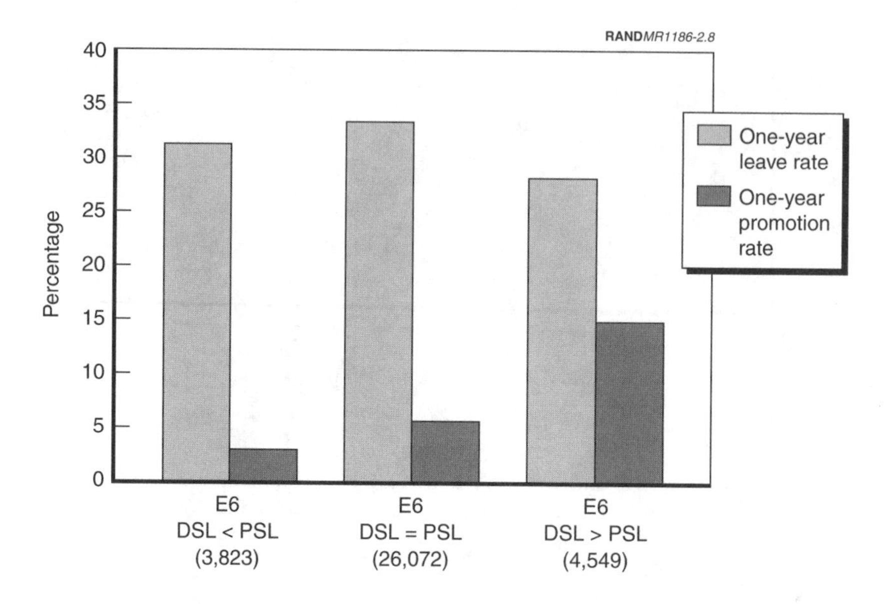

Figure 2.8—FY92 E6 One-Year Leave and Promotion Rates

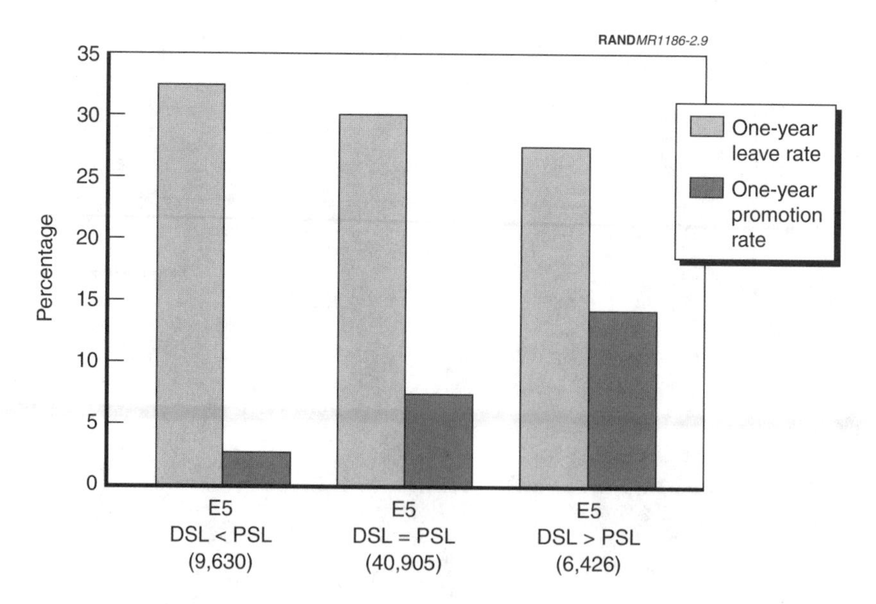

Figure 2.9—FY92 E5 One-Year Leave and Promotion Rates

Turning to retention rates, we note the low one-year retention rates for CMF 14, Air Defense Artillery. Fully half of the E6s in TOE units in November 1996 left the Army during the subsequent twelve months. CMF 74 (Automatic Data Processing), CMF 95 (Military Police), and CMF 98 (Signals Intelligence/Electronic Warfare) also have relatively low one-year retention rates (in the 55–70 percent range) as do several other CMFs. Similar tables for FY92 and FY94, E5s and E6s, can be found in Appendix A.[3]

In the next chapter, where we examine the implications of extending RCPs and improving mid-career retention, we will pay close attention to CMF 67, Aircraft Maintenance. This CMF was highlighted by EPMD NCOs as one of the hard-to-retain high-tech CMFs. Table 2.1 shows its one-year retention rates to be in the 70 percent range, and it has a substantial number of above-PSL E6s (142). Because it has more than 100 E6s, and because its above-PSL one-year retention rate is less than 75 percent, we chose to focus on this CMF when examining the effects of RCP extension and mid-career retention improvements.

Why Is There Upward Substitution?

Those serving above-PSL are being *upward substituted* into positions that require NCOs with a higher grade. What mechanisms are driving this upward substitution? Is it just the Army's desire to place its quality NCOs in more challenging assignments? This may be part of the reason. But there is another factor that contributes to upward substitution. Table 2.2 illustrates this.

The table lists all CMFs with more than 500 NCOs assigned to TOE units in November 1996. It shows the total number of NCOs so assigned (on the far right) and the percentage of assignments to authorized end strength for those units.

---

[3]In Table 2.1, CMF 23 (Air Defense System Maintenance) has some very low one-year retention rates. This appears to be an anomaly due to the fact that there are small numbers of NCOs in this MOS (which is why it is shaded in the table). Appendix A, which contains similar tables for FY92 and FY94, shows much higher one-year retention rates.

Table 2.1

FY96 E6 One-Year Promotion and Retention Rates

	Total NCOs in TOE Units			One-Year Promotion Rates (percent)			One-Year Retention Rates (percent)		
1996 E6	Below PSL	At PSL	Above PSL	Below PSL	At PSL	Above PSL	Below PSL	At PSL	Above PSL
11: Infantry	**484**	**3,094**	**322**	**1**	**4**	**19**	**66**	**63**	**66**
12: Combat Engineering	49	632	80	2	2	20	80	68	78
13: Field Artillery	251	1,856	245	1	5	25	70	68	70
14: Air Defense Artillery	81	452	98	2	6	15	42	50	46
18: Special Forces		887	312		20	23		83	79
19: Armor	**227**	**1,165**	**186**		**3**	**29**	**70**	**62**	**67**
23: Air Defense System Maintenance	20	40	13				20	13	8
25: Visual Information	13	28					62	64	
27: Land Combat and Air Defense System Maintenance	17	126	18		13	28	71	66	72
29: Signal Maintenance									
31: Signal Operations	354	1,249	117	1	5	23	66	64	77
33: Electronic Warfare/ Intercept Systems Maintenance	18	36	9				61	58	67
35: Electronic Maintenance and Calibration	52	259	4		3	25	60	62	100
37: Psychological Operations	23	38	11		8	27	61	58	55
46: Public Affairs	3	9	2		22		100	44	100
51: General Engineering	19	320	47		7	28	58	70	70

Table 2.1—Continued

	Total NCOs in TOE Units			One-Year Promotion Rates (percent)			One-Year Retention Rates (percent)		
1996 E6	Below PSL	At PSL	Above PSL	Below PSL	At PSL	Above PSL	Below PSL	At PSL	Above PSL
54: Chemical	130	615	101	2	5	16	75	76	83
55: Ammunition	68	386	36	1	7	33	82	74	86
63: Mechanical Maintenance	335	1,736	384	1	12	24	70	71	75
67: Aircraft Maintenance	315	1,227	142	2	4	25	71	69	73
71: Administration	179	662	173	3	7	15	65	65	73
74: Automatic Data Processing	56	228	46	4	9	9	55	59	63
77: Petroleum and Water	69	304	49	1	12	24	75	77	78
79: Recruiting and Reenlistment		19	1		16			63	100
81: Topographic Engineering	9	40	10		3	20	33	35	50
88: Transportation	236	1,336	126	1	5	20	71	73	74
91: Medical	305	1,039	132	2	5	16	64	70	61
92: Supply and Services	510	2,040	288	2	8	25	73	76	78
93: Aviation Operations	**39**	**109**	**27**	**5**	**8**	**15**	**72**	**65**	**81**
94: Food Services									
95: Military Police	162	893	124	2	2	8	67	70	60
96: Military Intelligence	**221**	**565**	**83**	**3**	**7**	**22**	**62**	**65**	**70**
97: Bands	120	358	60	5	6	22	85	87	92
98: Signals Intelligence/ Electronic Warfare Operations	**111**	**336**	**63**	**2**	**5**	**8**	**62**	**60**	**65**

All of the CMFs in Table 2.2 have assignments that fall below authorizations in all NCO grades. Table 2.3 shows those CMFs where assignments exceed authorizations in at least one grade. The same criteria apply here as were applied to Table 2.2, i.e., TOE units with more than 500 NCOs assigned in November 1996.

Comparing the CMFs in Table 2.3 with the above-, at-, and below-PSL promotion rates in Table 2.1 (the bold entries in Table 2.1), we see that, even in these CMFs, those E6s serving above-PSL have substantially higher one-year promotion rates than do their at- and below-PSL counterparts. Further, these CMFs have substantial proportions of below-PSL E6s. This implies that upward substitution is not driven solely by underassigned CMFs. Downward substitution also occurs in overassigned CMFs. Even in these CMFs, the Army

Table 2.2

Percentage of Assignments to Authorized Strength FY96 CMFs with Authorized Manning That Exceed Assignments

Career Management Field	E5	E6	E7	E8	Total Assigned
12: Combat Engineering	76%	93%	80%	88%	2,728
13: Field Artillery	89	98	88	76	7,472
18: Special Forces	0	86	89	88	2,918
27: Lnd Cmbt/Air Def Sys Maint	73	91	61	44	516
31: Signal Operations	87	91	81	84	7,825
35: Elec Maint and Calibration	63	65	62	59	1,438
51: General Engineering	87	95	75	79	1,569
54: Chemical	79	88	86	68	2,795
55: Ammunition	95	86	80	62	1,390
63: Mechanical Maintenance	85	91	79	69	11,975
67: Aircraft Maintenance	88	88	85	80	5,041
71: Administration	86	86	85	83	5,202
74: Automatic Data Processing	100	85	78	53	1,113
77: Petroleum and Water	87	98	81	59	2,169
88: Transportation	77	85	75	68	5,084
91: Medical	87	99	90	83	5,423
95: Military Police	82	97	92	71	3,874
97: Bands	77	92	93	86	1,310

assignment process is trying to place high-quality NCOs in challenging assignments.

Finally, how do the NCOES schooling attendance rates of at- and above-PSL NCOs compare? Figure 2.10 illustrates this, again for NCOs in TOE units in November 1996.

The three leftmost columns show percentages of above-PSL E4s, E5s, and E6s who have received the formal NCOES required by the positions they hold: PLDC for above-PSL E4s, BNCOC for above-PSL E5s, and ANCOC for above-PSL E6s. The rightmost three columns show the same percentages for those serving at-PSL, i.e., the percentages of at-PSL E5s, E6s, and E7s who have had the requisite NCOES schooling. Not surprisingly, most at-PSL NCOs have received the formal training for their positions (from 90 to 95 percent). With the exception of above-PSL E5s, most NCOs serving above-PSL have not received the formal NCOES schooling (from 70 to 75 percent). Forty percent of above-PSL E5s have not received formal NCOES schooling either. The above-PSL E4s, E5s, and E6s who have received formal NCOES schooling are probably on selection lists for the next-higher grade, but this leaves substantial proportions who are probably not on selection lists and therefore cannot receive the formal NCOES schooling. Relaxing select-train-promote to *select or assign*-train-promote would allow these NCOs to attend NCOES.

Table 2.3

Percentage of Assignments to Authorized Strength FY96 CMFs with Manning That Exceeds Authorization in at Least One Grade

Career Management Field	E5	E6	E7	E8	Total Assigned
11: Infantry	94%	105%	116%	80%	12,180
19: Armor	87	106	91	72	5,378
93: Aviation Operations	88	113	100	72	726
96: Military Intelligence	73	99	107	90	2,907
98: SIGINT/EW Operations	72	108	88	75	1,634

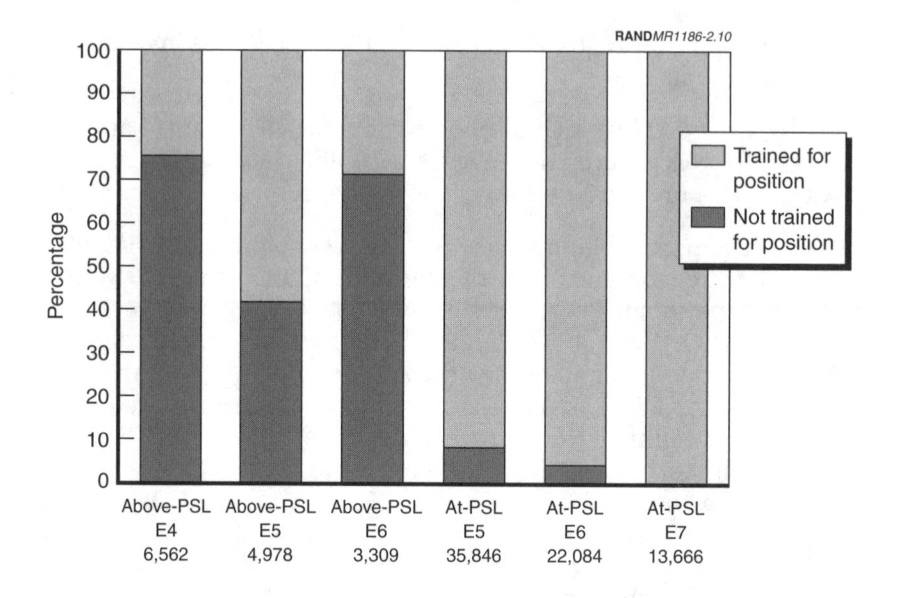

Figure 2.10—FY96 Above- and At-PSL NCOES Attendance Rates

Select-Train-Promote, NCOES, and Self-Development Implications

There clearly are NCOs who are serving above their primary skill levels, and the select-train-promote system is not allowing many of them to attend the NCOES school that would help them learn the technical and leadership skills they need to perform in these assignments. Select-train-promote was implemented solely to reduce NCOES costs. Relaxing select-train-promote to *select or assign*-train-promote, allowing those NCOs serving above-PSL to attend, would increase NCOES costs but would probably also enhance above-PSL NCO productivity. We believe the Army should move to *select or assign*-train-promote.

However, it may not be feasible to send all above-PSL NCOs to NCOES—their units may not be able to spare them. In this case it would be highly desirable to give above-PSL NCOs self-development assistance so that they can at least get exposure to those assignment-related skills that are amenable to self-development. The Army is currently investing heavily in distance learning, and making this capability available in a self-development environment would allow above-PSL NCOs to develop some of the skills needed in their assignments.

It would also be of benefit to the Army, once these above-PSL NCOs do get to NCOES, to provide a proficiency testing mechanism to determine which curriculum segments they have already mastered. Those who demonstrate mastery can be given enhanced skill training so that they don't have to relearn what they've already learned on the job. This would require some changes to the NCOES curriculum, but we believe the benefit to the Army would far exceed the cost of such changes.

Chapter Three

CHANGING MID-CAREER RETENTION AND RETENTION CONTROL POINTS

Attendees at the NCO Leader Development Workshop noted that NCOs in hard-to-retain, high-tech MOSs had to be forced out when they reached their RCPs, even though their skills were still in demand. To capitalize on the experience of these NCOs, the attendees recommended relaxing RCPs for these MOSs. Additionally, in meetings with senior EPMD NCOs, they noted that a large majority of NCOs who *crossed the 10-year boundary* stayed for 20 years. Finally, the above-, at-, and below-PSL retention analysis presented in Chapter Two indicates that substantial above-PSL NCOs are leaving the Army in the 7–10 year of service (YOS) period. These three issues motivate our examination of RCP relaxation and mid-career retention improvements. As we explained in Chapter Two, we've chosen CMF 67, Aircraft Maintenance, as the CMF on which to focus when considering these personnel policy changes. CMF 67 is a hard-to-retain high-tech CMF with sufficient numbers of above-PSL E6s in November 1996.

We examine the effects of relaxing E7 and E8 RCPs and of improving mid-career (years 7–10) E5 and E6 retention. Table 3.1 illustrates why these grades were chosen. It shows the inventory projection model's grade-YOS forecast for CMF 67.[1]

Improving mid-career (YOS 7–10) retention for E5s and E6s makes sense because these two grades have substantial numbers of NCOs in those years. The RCP for E4 is 10 years of service, and not too many (as a percentage of total E4s) make it to the RCP year. Further, there

[1]Appendix C presents the input parameters associated with CMF 67.

aren't very many E7s in years 7–10, and improving mid-career retention here won't have much of an effect on the force.

Extending RCPs for E7s and E8s makes sense because there are substantial numbers of E7s and E8s at their RCP years (as a percentage of total E7s and E8s). Further, not very many E3s, E4s, E5s, and E6s make it to their RCP years, thus making it unproductive to extend these RCPs.

Table 3.1

CMF 67 Steady-State Inventory by Grade and Year of Service

YOS	E1–3	E4	E5	E6	E7	E8	E9	Total
1	2,480							2,480
2	775	1,276						2,051
3	68	1,602	234					1,904
4	44	990	579					1,613
5	9	472	813	88				1,382
6	2	291	678	294				1,266
7		151	271	254				677
8		123	171	288				582
9		84	62	321	4			471
10		26	29	292	17			365
11			24	223	44			291
12			20	161	77			258
13			18	107	111			236
14			15	80	127			221
15			12	52	135	10		209
16				25	149	19		194
17				0	148	40		188
18				0	133	53	1	186
19				0	118	65	2	184
20				0	103	77	4	184
21					101	73	7	181
22					5	38	8	52
23						26	10	35
24						17	8	25
25							8	8
26							6	6
27							5	5
28							4	4
29							2	2
30							2	2
Total	3,377	5,016	2,927	2,187	1,271	418	66	15,262

Improve retention here

Extend RCPS here

In extending E7 and E8 RCPs, we allow 50 percent to extend three years beyond the RCP year. The rationale here is that we don't expect all E7s and E8s who make it to their RCP year to want to extend (or the Army may not want all E7s and E8s in CMF 67 to extend beyond their RCP year). In each of the extension years we also impose 10 percent attrition as a reasonable estimate of leakage during those years.

For completeness, we also consider an RCP extension case where 90 percent of E7s and E8s are allowed to go beyond the RCP year, also with 10 percent leakage during the three extension years. This case, while probably unrealistic, is included to provide an upper bound on the experience benefits available through the RCP extension alternative.

We also consider two mid-career retention improvement cases for the grades of E5 and E6: 25 percent improvement and 50 percent improvement (which translates to 25 percent and 50 percent reductions in year 7–10 attrition rates).

E7 AND E8 RCP RELAXATION

Relaxing E7 and E8 RCPs means that either 50 or 90 percent of the E7s and E8s who reach their RCP years will be allowed to stay for three additional years (with a 10 percent leakage in each of the first two years). Figure 3.1 shows how such relaxation affects the E7 YOS profile. Figure 3.2 shows the same for the E8 YOS profile.

Each curve in the two figures shows a distinct steady-state YOS profile. In Figure 3.1 the number of E7s is the same for each curve: 1,271 in the inventory. In Figure 3.2 the number of E8s is also the same for each curve: 418 in the inventory. The base case shows the NCOs leaving in the RCP year: YOS 21 in the case of E7s, and YOS 24 for E8s. The 50 percent RCP relaxation case shows 50 percent of the NCOs who reach this RCP year extending for three more years, with a 10 percent leakage in the first two extension years—this leakage is why the post-RCP segments are not horizontal. The 90 percent relaxation case shows 90 percent of the NCOs who reach this RCP year extending for three more years, also with a 10 percent leakage in the first two extension years.

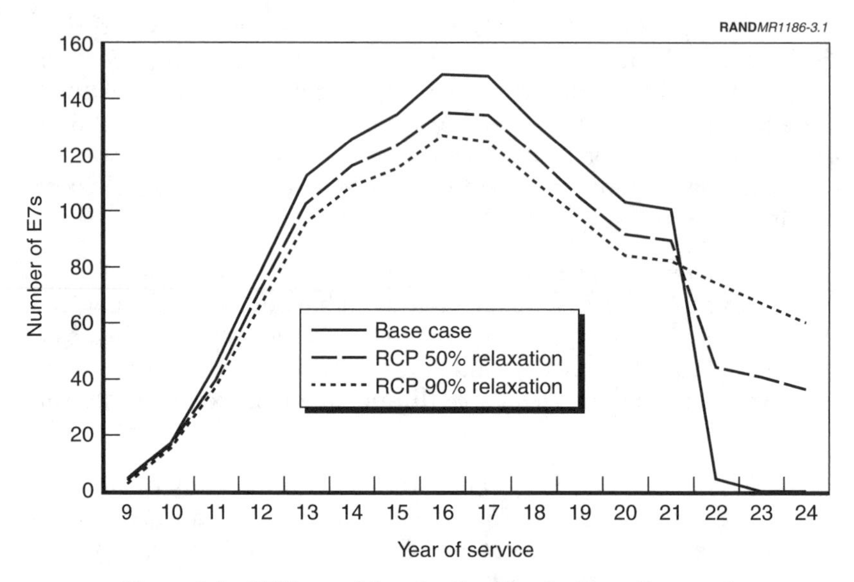

Figure 3.1—E7 Year-of-Service Profiles for Base Case and RCP Relaxation Cases

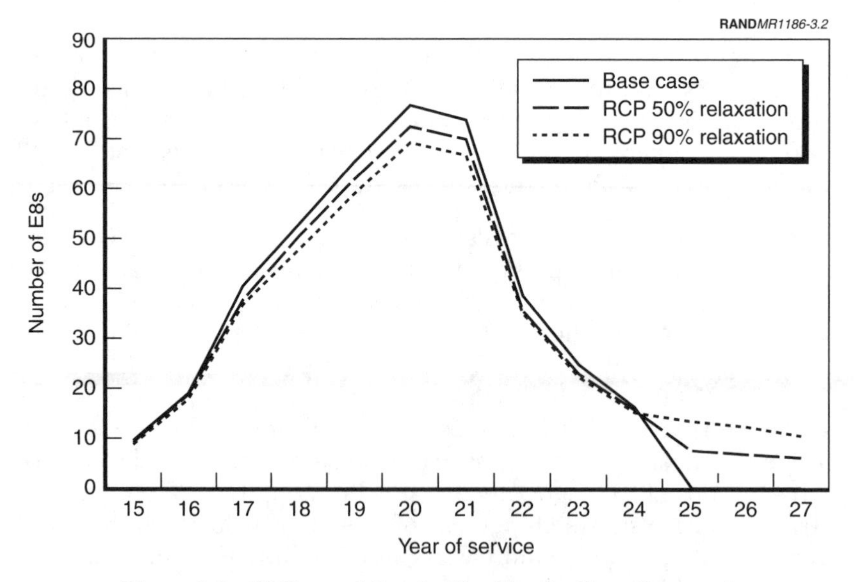

Figure 3.2—E8 Year-of-Service Profiles for Base Case and RCP Relaxation Cases

Focusing on Figure 3.1, the 50 percent relaxation case shows about 120 E7s extending beyond the RCP, or about 10 percent of the E7 inventory. The 90 percent relaxation case shows about 200 E7s extending beyond the RCP, or about 16 percent. This means that the number of E7s in pre-RCP years must be reduced by 120 or 200. Hence the base case curve stands above the RCP relaxation curves in the pre-RCP years and falls below in the post-RCP years. Stated differently, in order to benefit from the experience of 120 (or 200) post-RCP E7s, we must reduce the pre-RCP E7s by 120 (or 200). While this increases the overall experience of E7s, it also reduces the promotion chances of E6s competing for promotion to E7 by 5 (or 9) percent. This may result in a small drop in E6 retention, but we did not try to capture this because there are comparable increases in promotion chances to E8 and E9.

Figure 3.2 shows the same pattern for E8s. In the 50 percent relaxation case, the number of post-RCP E8s is 21, or about 5 percent of the E8 force. In the 90 percent case this number is 37, or about 9 percent.

RCP Relaxation Force Characteristic Comparisons

Three force characteristics are used as measures of effectiveness: the number of NCOs (E5 and above) with 11 or more years of service; the average time in each grade; and the probability of promotion to each grade.[2] Table 3.2 presents these force characteristics for the base case and two RCP relaxation cases.

The first force characteristic (NCOs with 11 or more years of service) has some surprising effects. Relaxing E7 and E8 RCPs has no effect on the number of E7s and E8s with 11 or more years of service, because most E7s and all E8s have 11 or more years of service. However, the numbers of E5s and E6s with 11 or more years of ser-

[2]The average time in grade in the steady state is the ratio of the total number of NCOs in that grade and the number of promotions into the grade. The probability of promotion to a grade **g** is defined to be the probability, given that an NCO was just promoted to grade **g – 1**, that he will be promoted to grade **g** sometime during his career. This turns out to be the ratio of the number of promotions to grade **g** and the number of promotions into grade **g – 1**.

Table 3.2

RCP Relaxation Force Characteristics Compared with Base Case

	E1–3	E4	E5	E6	E7	E8	E9	Total
CMF 67 NCOs with 11 or more years of service								
Base case			89	649	1,250	418	66	2,472
50% E7/E8 RCP relaxation			101	698	1,252	418	66	2,535
90% E7/E8 RCP relaxation			110	731	1,253	418	66	2,579
% difference: 50% relaxation			+14%	+7%	+0%			+3%
% difference: 90% relaxation			+24%	+13%	+0%			+4%
CMF 67 average time in grade								
Base case	1.4	2.6	2.3	4.2	5.8	5.1	6.1	
50% E7/E8 RCP relaxation	1.4	2.7	2.3	4.3	6.3	5.3	6.3	
90% E7/E8 RCP relaxation	1.4	2.7	2.4	4.4	6.7	5.5	6.3	
% difference: 50% relaxation	+0%	+1%	+1%	+3%	+9%	+5%	+4%	
% difference: 90% relaxation	+1%	+1%	+2%	+6%	+16%	+9%	+4%	
CMF 67 probability of promotion into grade								
Base case		0.77	0.67	0.41	0.42	0.37	0.13	
50% E7/E8 RCP relaxation		0.77	0.66	0.41	0.39	0.39	0.13	
90% E7/E8 RCP relaxation		0.76	0.66	0.40	0.38	0.40	0.14	
% difference: 50% relaxation		–0%	–1%	–2%	–5%	+4%	+1%	
% difference: 90% relaxation		–0%	–1%	–3%	–9%	+7%	+4%	

vice increase substantially. This occurs because the number of promotions to E7 (and therefore to E6) is reduced when E7 and E8 RCPs are extended. The E6s (and E5s) who would have been promoted will serve in grade for longer periods, subject to the normal attrition for those grades. This effect acts to increase the numbers of E5s and E6s with more than 10 years of service.

Figures 3.3 and 3.4 show the E5 and E6 YOS profiles for the base case and the two RCP relaxation cases. While the force characteristics above show substantial increases in E5s and E6s with more than 10 years of service, the two figures show that, based on the entire E5 and E6 populations, these increases do not have a major effect on the E5 and E6 YOS profiles.

Returning to Table 3.2, the second force characteristic is average time in grade. Not surprisingly, E7 and E8 average times in grade increase from 5 to 16 percent. These increases are driven by the fact that promotions into E7 and E8 are reduced in the RCP relaxation cases. This has a ripple effect on the lower grades, with their average times in grade also increasing. The increases are not large, and the effect diminishes as we move from the higher to the lower grades: 3 to 6 percent for E6, down to an almost undetectable 0 to 1 percent for E1–3.[3]

The third force characteristic is probability of promotion. Relaxing E7/E8 RCPs leads to reduced promotions into E7. These result in a 5 to 9 percent reduction in promotion probability to E7. As with average time in grade, there is a ripple effect on the lower grades, with reductions in promotion probabilities, but at a steadily decreasing rate. Interestingly, promotion probability to E8 increases. The E7s

---

[3]E9 average time in grade also increases, but this is more a modeling artifact than a result of RCP relaxation. The model works from high grade to low grade, determining the year of service profile and promotion requirements for E9 before doing the same for E8, before E7, and so forth. In the base case there are sufficient promotion-eligible E8s to support E9 promotion needs. But when E8 RCPs are relaxed, the number of E8s is spread more sparsely over the years of service, leading to a situation where E9 promotions in at least one year of service cannot be supported by the promotion-eligible E8s. The solution to this is to alter the E9 promotion profile, leading to earlier (and fewer) promotions into E9 and therefore the modest increase in E9 average time in grade.

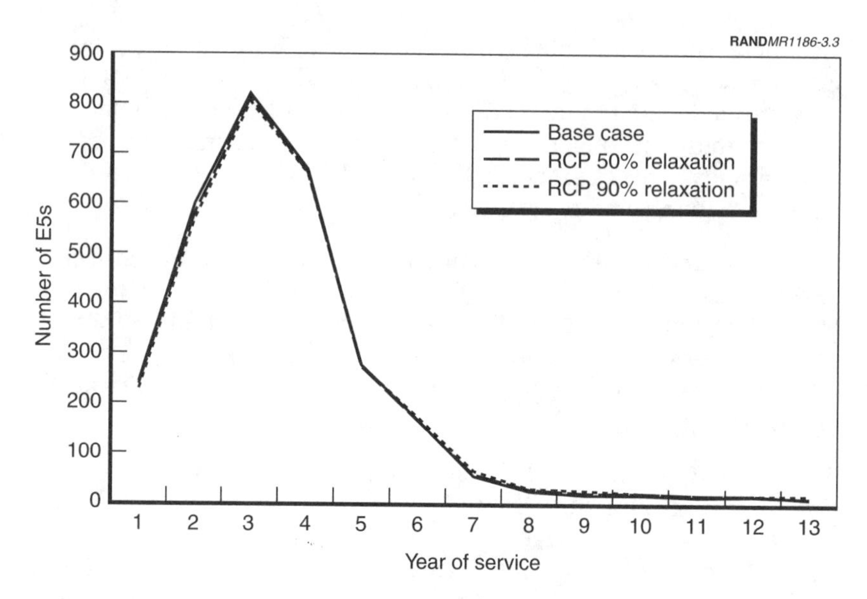

Figure 3.3—E5 Year-of-Service Profiles for Base Case and RCP Relaxation Cases

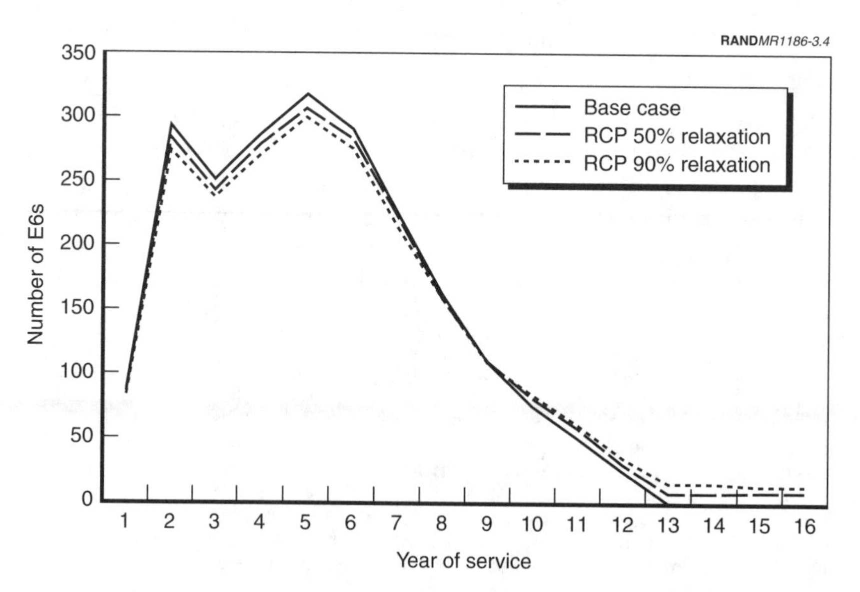

Figure 3.4—E6 Year-of-Service Profiles for Base Case and RCP Relaxation Cases

have more opportunities to get promoted to E8 when RCPs are relaxed. The same holds true for promotions to E9.[4]

E5 AND E6 MID-CAREER RETENTION IMPROVEMENT

We examined two mid-career retention improvement cases: improving E5/E6 YOS 7–10 retention by 25 percent, and improving it by 50 percent. These grades and year ranges were chosen because of statements by EPMD NCOs to the effect that soldiers who are in the force at the 10-year point stand a very high chance of being in the force at the 20-year point. Given the force profile illustrated in Table 3.1, E5s and E6s are the only reasonable grades where mid-career retention improvements make sense.

Figures 3.5 and 3.6 show the YOS profiles for the grades of E5 and E6 in the base case and two retention improvement cases. The E5 profiles show the base case curve higher than the other two curves in years 3 through 7. The base case curve falls below the retention improvement curves in years 9 through 15. For E7 profiles, the years are 5–9 and 11–20.

Table 3.3 presents the force characteristics for the two retention improvement cases, along with the base case and the RCP relaxation cases. First, E5 and E6 mid-career retention improvement, be it 25 or 50 percent, has a much greater positive effect on force characteristics than does E7 and E8 RCP relaxation. Whereas RCP relaxation results in a 3 to 4 percent increase in the numbers of NCOs with more than ten years of service, mid-career retention improvement increases this by 6 to 12 percent.

Second, average times in grade for E5s and E6s also show more dramatic increases than do the RCP relaxation cases: 5 to 17 percent versus 1 to 6 percent. However, relaxing E7 and E8 RCPs does lead to increases in E7 and E8 average times in grade: from 5 (E8) to 16 (E7) percent.

---

[4]The quantitative reason for increased promotion probability to E8 arises because promotions into E8 drop at a slower rate than do promotions into E7, and probability of promotion is the ratio of E8 promotions to E7 promotions. Similarly, promotions into E9 drop at a slower rate than do promotions into E8, leading to increased promotion probability to E9.

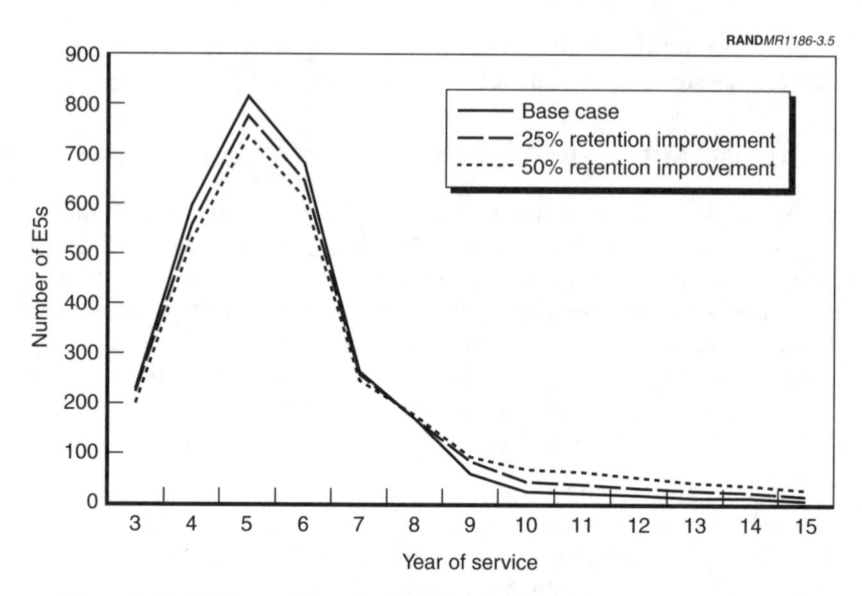

Figure 3.5—E5 Year-of-Service Profiles for Base Case and Mid-Career Retention Improvement Cases

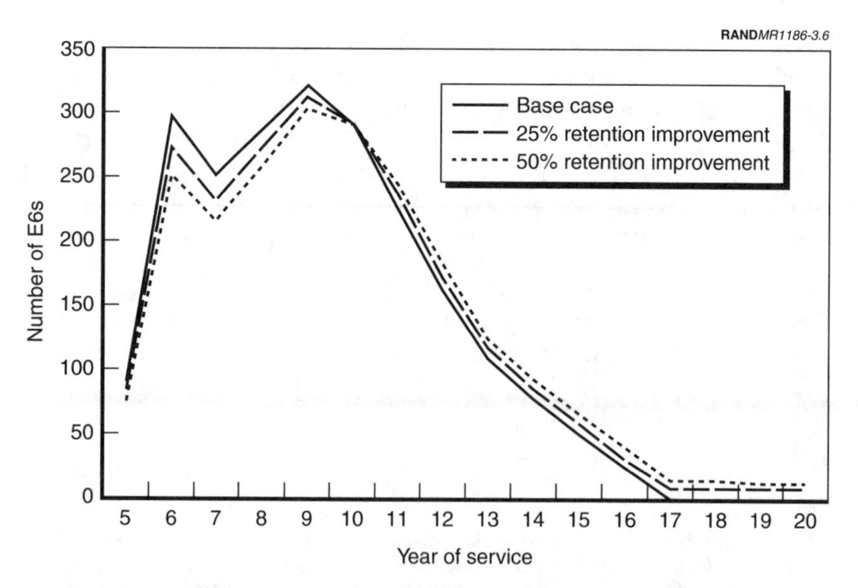

Figure 3.6—E6 Year-of-Service Profiles for Base Case and Mid-Career Retention Improvement Cases

Finally, improving E5 and E6 mid-career retention, while keeping the inventories constant, results in modest drops in promotion probability to E5 and E6 (2 to 6 percent). However, promotion probability to E7 increases from 8 to 17 percent. The E5 and E6 promotion probability drops because E5s and E6s stay longer, thereby reducing the number of promotions necessary to sustain the E5 and E6 inventories. Figures 3.5 and 3.6 illustrate this through changes in the E5 and E6 YOS profiles toward older E5 and E6 inventories. E7 promotion probability increases because the number of E7 promotions doesn't change while the number of E6 promotions declines.

COMPARING RCP RELAXATION WITH MID-CAREER RETENTION IMPROVEMENT

Table 3.3 permits ready comparison of E7/E8 RCP relaxation and E5/E6 mid-career retention improvement. Both E7/E8 RCP relaxation and E5/E6 mid-career retention improvement lead to improvements in NCO experience levels, with only modest reductions in promotion probabilities and some senior grade promotion probability increases. But when we compare the two types of change, it is clear that improving E5/E6 mid-career retention has a far greater effect on NCO experience levels than does the extension of E7/E8 RCPs. Even when we allow most E7s and E8s to remain in the Army three years beyond their RCPs, the effect on YOS 11 and above experience is much less than the increased experience levels provided by improving E5 and E6 mid-career retention.

However, there is another side to this issue. In implementing a targeted E7/E8 RCP relaxation policy, the Army may encounter problems. While it would be relatively easy for the Army to selectively extend E7 and/or E8 RCPs for specific MOSs, it is not clear how many takers the Army would get. There may need to be some inducement for targeted E7s and E8s to extend. Further, in implementing a targeted E5/E6 mid-career retention improvement policy, the Army would have to provide some inducement to targeted E5s and E6s to remain in the Army, sufficient to get them to stay past their tenth year. Finally, if such inducements are successful, thereby leading to a more experienced NCO corps, there will be increased pay and retirement costs as well as reduced accession and training costs. We

Table 3.3

Retention Improvement Force Characteristics Compared with the Base Case and the RCP Relaxation Cases

CMF 67 NCOs with 11 or more years of service	E1–3	E4	E5	E6	E7	E8	E9	Total
Base case			89	649	1,250	418	66	2,472
50% E7/E8 RCP relaxation			101	698	1,252	418	66	2,535
90% E7/E8 RCP relaxation			110	731	1,253	418	66	2,579
% difference: 50% relaxation			+14%	+7%				+3%
% difference: 90% relaxation			+24%	+13%				+4%
25% E5/E6 retention improvement			158	727	1,250	418	66	2,619
50% E5/E6 retention improvement			238	796	1,250	418	66	2,768
% difference: 25% improvement			+79%	+12%				+6%
% difference: 50% improvement			+169%	+23%				+12%
CMF 67 average time in grade	E1–3	E4	E5	E6	E7	E8	E9	Total
Base case	1.4	2.6	2.3	4.2	5.8	5.1	6.1	
50% E7/E8 RCP relaxation	1.4	2.7	2.3	4.3	6.3	5.3	6.3	
90% E7/E8 RCP relaxation	1.4	2.7	2.4	4.4	6.7	5.5	6.3	
% difference: 50% relaxation	+0%	+1%	+1%	+3%	+9%	+5%	+4%	
% difference: 90% relaxation	+1%	+1%	+2%	+6%	+16%	+9%	+4%	
25% E5/E6 retention improvement	1.4	2.7	2.4	4.5	5.8	5.1	6.1	
50% E5/E6 retention improvement	1.4	2.8	2.6	4.9	5.8	5.1	6.1	
% difference: 25% improvement	+1%	+2%	+5%	+8%				
% difference: 50% improvement	+3%	+5%	+12%	+17%				

Table 3.3—continued

CMF 67 probability of promotion into grade	E1–3	E4	E5	E6	E7	E8	E9	Total
Base case		0.77	0.67	0.41	0.42	0.37	0.13	
50% E7/E8 RCP relaxation		0.77	0.66	0.41	0.39	0.39	0.13	
90% E7/E8 RCP relaxation		0.76	0.66	0.40	0.38	0.40	0.14	
% difference: 50% relaxation		–0%	–1%	–2%	–5%	+4%	+1%	
% difference: 90% relaxation		–0%	–1%	–3%	–9%	+7%	+4%	
25% E5/E6 retention improvement		0.76	0.65	0.40	0.45	0.37	0.13	
50% E5/E6 retention improvement		0.75	0.63	0.40	0.49	0.37	0.13	
% difference: 25% improvement		–1%	–3%	–2%	+8%			
% difference: 50% improvement		–2%	–6%	–5%	+17%			

have not attempted to estimate any of these costs, and the overall utility of the two policy alternatives depends on these costs as well as the experience benefits the alternatives afford.

Chapter Four

SUMMARY OF RESEARCH FINDINGS AND RECOMMENDATIONS

This chapter reviews the NCO leader development project's major retention findings and the recommendations that emerge from those findings.

ABOVE-PSL FAST TRACKERS

NCOs serving above their primary skill level are both fast trackers and just as likely to leave the Army as are their slower tracking at- and below-PSL counterparts. The one-year promotion rates for above-PSL NCOs, and especially E6s, are remarkably higher than those for their at- and below-PSL counterparts. The Army is clearly placing NCOs who show promise in positions that normally require a higher grade, even though there are NCOs of the higher grade who are serving in below-PSL assignments.

On a CMF basis, the one-year promotion rates in November 1996 for above-PSL E6s can be anywhere from 2 to 30 times higher than below-PSL E6s of the same CMF. The majority of these CMFs have above-PSL promotion rates that are 20–25 times higher than their below-PSL counterparts—see Table A.6 in Appendix A. Yet above-PSL E6s have one-year leave rates that are very close to their at- and below-PSL counterparts—in the 30 percent range.

This raises the question as to why above-PSL NCOs are being allowed to leave at such high rates. Indeed, can the Army target above-PSL mid-career (years 7 through 10) E5s and E6s in hard-to-retain CMFs with special inducements to remain in the Army? If these NCOs can be induced to remain in the Army through their tenth year of service,

Army experience shows that a large majority of them will stay to at least twenty years of service. For CMF 67 (Aircraft Maintenance), improving mid-career E5/E6 retention by 50 percent leads to a 12 percent increase in the number of NCOs with more than ten years of service, coupled with a 2 to 5 percent reduction in junior NCO promotion probability (Table 3.3).

What is causing the Army to assign NCOs to above-PSL positions? Looking at the ratio of assignments to authorizations for each NCO grade within a CMF, we find that most CMFs are undermanned, i.e., they have fewer duty-MOS-qualified NCOs in a grade than the authorized end strength for that CMF/grade requires. These undermanned CMFs require NCOs to fill these positions, so there is a natural tendency to upwardly substitute lower-grade NCOs. However, even in these CMFs there are NCOs serving below-PSL (comparing Tables 2.1 and 2.2 shows this to be a common occurrence).

SELECT OR ASSIGN-TRAIN-PROMOTE FOR ABOVE-PSL FAST TRACKERS

The Army's select-train-promote NCOES attendance policy does not permit an above-PSL NCO to attend the NCOES associated with the higher grade *unless the NCO has been selected for the higher grade.* There are above-PSL NCOs who have not as yet been selected, and these soldiers are required to develop on the job the skills they need for their assignments. Relaxing select-train-promote for these NCOs to *select or assign*-train-promote would allow them to attend the formal NCOES school in a timely manner and perform more effectively in the above-PSL assignment.[1]

IMPROVED SELF-DEVELOPMENT OPPORTUNITIES FOR ABOVE-PSL FAST TRACKERS

Even if select-train-promote is relaxed to *select or assign*-train-promote, not all above-PSL NCOs would be able to attend formal

[1]It is possible that some of the above-PSL NCOs who would be allowed to attend NCOES may still fail of promotion, thereby "wasting" some school slots. However, even if they fail of promotion, they would still have the formal skills to better perform in their above-PSL assignment.

NCOES. For example, the NCO's unit may not be able to spare him. In these situations it makes sense to provide self-development facilities geared to instruct/expose those skills that are amenable to self-development. The Army today is in the process of expanding self-development facilities, including the acquisition of distance learning capabilities. Using above-PSL needs can provide focus to this acquisition effort.

UP-FRONT SKILLS ASSESSMENT AND ENHANCED CURRICULA AT NCOES

NCOs who have served above-PSL, when they do attend NCOES, have already learned on the job the skills that the NCOES course is designed to teach. For these NCOs it makes sense to provide up-front proficiency testing when they arrive at NCOES. NCOs who can demonstrate proficiency in the skills can better utilize their NCOES time if they are provided with supplemental enhanced curriculum instead of instruction in the skills for which they have demonstrated proficiency. This enhanced curriculum would make these NCOs more valuable to the Army. Since these NCOs are fast trackers, it makes sense for the Army to expose them to the enhanced curriculum.[2]

Providing the ability to test entering NCOES students for proficiency is probably not a major issue. The NCOES already tests for proficiency after NCOs complete the formal curriculum. However, providing enhanced curriculum for those entering NCOs who have demonstrated proficiency may require important multiple-track adjustments in the way NCOES currently does business.

RETENTION ISSUES

The preceding discussion focuses primarily on taking steps to help above-PSL NCOs better perform in their assignments. It does not dwell on the retention behavior of E5s and E6s or the RCP force-out

---

[2]Some NCO Leadership Workshop attendees noted that formal NCOES instruction in skills already learned on the job has advantages. It provides those NCOs with a mechanism for validating and adjusting their on-the-job skills as a result of being exposed to the formal curriculum.

of E7s and E8s. Yet improving above-PSL E5 and E6 retention behavior and selectively extending E7 and E8 RCPs can improve the NCO corps' experience mix.

CMF 67 (Aircraft Maintenance) was selected for this analysis. CMF 67 is a hard-to-retain CMF, is relatively high-tech, and had more than 100 above-PSL E6s assigned to TOE units in November 1996.

E7/E8 RCP Relaxation

Allowing 50 percent of E7s and E8s to serve three years beyond their RCPs increases by 3 percent the number of NCOs with more than ten years of service (see Table 3.3). Allowing 90 percent to serve three years beyond their RCPs shows a 4 percent increase. But if we focus on E5s and E6s, selectively extending E7 and E8 RCPs by three years increases by 7 to 24 percent the numbers of E5s and E6s with more than ten years. This increase is driven by the fact that fewer E5s and E6s will be promoted, and they will thus spend more time in E5 and E6. These additional E5s and E6s therefore are there because they failed to be promoted to E6 and E7 respectively.

E5/E6 Mid-Career Retention Improvement

If, instead of extending E7/E8 RCPs, we try to retain mid-career E5s and E6s, there is a more dramatic increase in NCO experience levels. Improving year 7–10 E5 and E6 retention by 25 percent yields a 6 percent increase in NCO experience (numbers of NCOs with more than ten years of service). A 50 percent retention improvement doubles this to 12 percent. This improvement comes from above-PSL E5s and E6s.

Improving mid-career E5/E6 retention not only has a quantitative edge over extending E7/E8 RCPs, it has a qualitative edge as well. Where the E7/E8 RCP extension improvement shows an increase in the number of senior E5s and E6s, this increase comes from E5s and E6s who have failed of promotion. The mid-career retention improvement, on the other hand, retains above-PSL, fast-tracker E5s and E6s, not E5s and E6s who have failed of promotion.

COST CONCERNS

Both improving E5/E6 mid-career retention and relaxing E7/E8 RCPs will lead to a more senior force. This enhanced seniority will increase compensation and retirement costs. Further, to improve mid-career retention of fast-tracking E5s and E6s in high-tech and hard-to-retain MOSs will require additional incentives, and this means added costs. Incentives may take the form of enhanced education opportunities and improved self-development capabilities. They may also take the form of financial incentives targeted at fast trackers to induce them to stay in the Army. These cost increases will be mitigated somewhat by reductions in accession and training costs. This analysis has not addressed these cost issues.

Appendix A

FY92, FY94, AND FY96 E5 AND E6 ONE-YEAR PROMOTION AND RETENTION RATES BY CAREER MANAGEMENT FIELD

This appendix contains six tables, showing for each CMF the one-year promotion and retention rates for E5s and E6s in fiscal years 1992, 1994, and 1996. The rates are based on the numbers serving in TOE units in November of the indicated year, and these numbers are included in the tables as well. Shaded rows in the tables highlight those CMFs with fewer than 100 NCOs in the given grade.

Two columns of asterisks are included to the right of the one-year promotion rates and one-year retention rates. An asterisk in the left column means that the below-PSL rate is greater than the at-PSL rate. An asterisk in the right column means that the at-PSL rate is greater than the above-PSL rate.

Table A.1

FY92 E5 One-Year Promotion and Retention Rates by CMF

	Total NCOs			One-Year Promotion Rates (percent)				One-Year Retention Rates (percent)				
1992 E5	Below PSL	At PSL	Above PSL	Below PSL	At PSL	Above PSL		Below PSL	At PSL	Above PSL		
11: Infantry	800	4,451	888	4	9	24		72	73	78		
12: Combat Engineering	127	1,219	319	5	15	29		71	79	84		
13: Field Artillery	561	3,055	636	3	10	19		81	80	85	*	
14: Air Defense Artillery	162	695	138	2	6	1	*	40	38	28	*	*
18: Special Forces		8	71		88	80	*		88	97		
19: Armor	259	2,293	364	1	3	4		75	79	85		
23: Air Defense System Maintenance	26	125	14	4	14	14	*	46	52	57		
25: Visual Information	19	27	3		7		*	58	78	100		
27: Land Combat and Air Defense System Maintenance	46	258	19	2	10	21		76	78	79		
29: Signal Maintenance	175	635	81	3	8	11		70	69	79	*	
31: Signal Operations	503	3,108	249	2	7	11		59	59	80		
33: Electronic Warfare/ Intercept Systems Maintenance	33	57	7	3	14	29		82	75	71	*	*
35: Electronic Maintenance and Calibration	33	172	24	3	3		*	48	61	50		*
37: Psychological Operations	24	45	4	4	20	50		83	67	100	*	
46: Public Affairs	11	9	2					27	33	50		
51: General Engineering	209	953	21	4	7	5	*	72	74	76		
54: Chemical	126	892	211	4	9	12		81	80	86	*	

Table A.1—continued

1992 E5	Total NCOs Below PSL	Total NCOs At PSL	Total NCOs Above PSL	One-Year Promotion Rates (percent) Below PSL	One-Year Promotion Rates (percent) At PSL	One-Year Promotion Rates (percent) Above PSL			One-Year Retention Rates (percent) Below PSL	One-Year Retention Rates (percent) At PSL	One-Year Retention Rates (percent) Above PSL		
55: Ammunition	70	616	74	4	4	16	*		77	71	73	*	
63: Mechanical Maintenance	1,543	5,655	449	1	5	11			79	78	79	*	
67: Aircraft Maintenance	759	1,731	234	4	13	18			75	75	75	*	
71: Administration	707	1,950	488	1	5	8			65	69	73		
74: Automatic Data Processing	32	312	101		0	2			66	59	68	*	
77: Petroleum and Water	278	733	57	4	15	7		*	82	82	79		*
79: Recruiting and Reenlistment													
81: Topographic Engineering	29	104	5		6	20			76	78	80		
88: Transportation	569	2,094	249	1	7	12			75	79	85		
91: Medical	527	2,413	296	3	10	14			59	72	72		*
92: Supply and Services	873	2,729	595	1	2	3			27	23	21	*	*
93: Aviation Operations	96	189	54	8	16	20			70	69	80	*	
94: Food Services	290	1,021	196	7	13	17			77	81	83		
95: Military Police	282	1,695	208		1	3			64	68	75		
96: Military Intelligence	153	765	153	6	10	15			66	72	73		
97: Bands	109	197	114	6	6	12	*		86	83	81	*	*
98: Signals Intelligence/Electronic Warfare Operations	182	693	102	2	8	13			61	64	75		

Table A.2

FY92 E6 One-Year Promotion and Rhetention Rates by CMF

	Total NCOs			One-Year Promotion Rates (percent)			One-Year Retention Rates (percent)				
1992 E6	Below PSL	At PSL	Above PSL	Below PSL	At PSL	Above PSL	Below PSL	At PSL	Above PSL		
11: Infantry	386	3,484	577	2	7	17	66	69	74		
12: Combat Engineering	65	678	160	6	10	25	78	76	81	*	
13: Field Artillery	218	2,110	353	2	3	16	83	79	86	*	
14: Air Defense Artillery	70	613	63	1	2	5	44	39	33	*	*
18: Special Forces	5	709	259		17	28	100	84	86	*	
19: Armor	58	1,387	262	3	5	16	66	75	80		
23: Air Defense System Maintenance	28	81	19	4	5	26	54	51	58	*	
25: Visual Information	6	12					50	67			*
27: Land Combat and Air Defense System Maintenance	28	197	11		2	9	64	70	82		
29: Signal Maintenance	76	299	20		3	10	62	67	45		*
31: Signal Operations	162	1,603	163	2	4	5	71	51	31	*	*
33: Electronic Warfare/ Intercept Systems Maintenance	27	41	4				59	63	50		*
35: Electronic Maintenance and Calibration	39	123	20		2	10	69	52	45	*	*
37: Psychological Operations	21	16	7		25	29	71	75	71		*
46: Public Affairs	4	11	9			33	75	45	78	*	
51: General Engineering	9	379	48		5	23	67	82	73		*
54: Chemical	115	722	163	3	10	18	84	82	83	*	

Table A.2—continued

1992 E6	Total NCOs			One-Year Promotion Rates (percent)					One-Year Retention Rates (percent)				
	Below PSL	At PSL	Above PSL	Below PSL	At PSL	Above PSL			Below PSL	At PSL	Above PSL		
55: Ammunition	36	390	72		1	1			75	63	69	*	
63: Mechanical Maintenance	246	2,329	476	4	8	14			82	77	81	*	
67: Aircraft Maintenance	398	1,130	135	3	4	13			76	75	86	*	
71: Administration	210	1,048	171	4	8	14			58	62	71		
74: Automatic Data Processing	29	289	60	14	1	3	*		66	58	65	*	
77: Petroleum and Water	40	190	56	10	14	38			85	75	79	*	
79: Recruiting and Reenlistment	1	51	9		18	11		*	100	65	44	*	*
81: Topographic Engineering	11	52	8	9	8		*	*	55	75	88		
88: Transportation	189	1,395	189	1	4	11			76	81	85		
91: Medical	357	1,204	231	3	7	10			65	67	71		
92: Supply and Services	241	2,381	318	2	2	4			23	22	20	*	*
93: Aviation Operations	48	173	52	2	5	15			73	58	79	*	
94: Food Services	101	748	155	10	14	20			87	81	83	*	
95: Military Police	154	1,002	167	4	1	10	*		66	67	61		*
96: Military Intelligence	176	505	123	1	6	14			70	71	77		
97: Bands	161	412	99	3	6	15			89	90	94		
98: Signals Intelligence/Electronic Warfare Operations	103	306	90	5	6	19			71	67	77	*	

Table A.3

FY94 E5 One-Year Promotion and Retention Rates by CMF

	Total NCOs			One0Year Promotion Rates (percent)					One-Year Retention Rates (percent)				
1994 E5	Below PSL	At PSL	Above PSL	Below PSL	At PSL	Above PSL			Below PSL	At PSL	Above PSL		
11: Infantry	820	4,760	989	2	3	7			68	72	70		*
12: Combat Engineering	151	1,454	236	9	11	24			72	75	76		
13: Field Artillery	420	3,602	663	3	3	4			76	77	78		
14: Air Defense Artillery	160	666	80	3	5	18			60	62	64		
18: Special Forces	1	23	133	100	83	78	*	*	100	87	90	*	
19: Armor	338	2,550	346	9	11	26			74	72	75	*	
23: Air Defense System Maintenance	40	93	13	8	12	8		*	80	46	62	*	
25: Visual Information	3	10	2		10			*	67	40	50	*	
27: Land Combat and Air Defense System Maintenance	55	345	18	2	2			*	58	72	33		*
29: Signal Maintenance	177	528	52	1	2	2		*	14	16	17		
31: Signal Operations	603	3,275	450	2	5	4		*	52	49	53	*	
33: Electronic Warfare/ Intercept Systems Maintenance	47	63	1	11	11			*	64	63	100	*	
35: Electronic Maintenance and Calibration	27	165	16		7	25			67	61	63	*	
37: Psychological Operations	27	45	12	11	20	8		*	81	87	92		
46: Public Affairs	14	10	9		10	11			21	60	78		
51: General Engineering	90	915	9	4	8	11			61	62	67		
54: Chemical	106	1,088	237	6	6	10			75	76	78		

Table A.3—continued

	Total NCOs			One0Year Promotion Rates (percent)					One-Year Retention Rates (percent)				
1994 E5	Below PSL	At PSL	Above PSL	Below PSL	At PSL	Above PSL			Below PSL	At PSL	Above PSL		
55: Ammunition	77	539	59	3	5	15			77	77	73		*
63: Mechanical Maintenance	1,201	6,470	477	4	5	8			75	73	76	*	
67: Aircraft Maintenance	672	2,146	272	7	8	11			75	71	68	*	*
71: Administration	566	2,074	491	6	8	12			60	59	63	*	
74: Automatic Data Processing	66	324	112	17	9	6	*	*	58	48	34	*	*
77: Petroleum and Water	338	1,003	72	6	10	19			81	79	72	*	*
79: Recruiting and Reenlistment													
81: Topographic Engineering	35	137	5	9	7	20	*		66	33	40	*	
88: Transportation	440	2,408	268	1	5	10			77	78	82		
91: Medical	425	2,657	246	1	5	9			52	57	65		
92: Supply and Services	624	2,914	575	3	6	11			65	71	70		*
93: Aviation Operations	122	246	35	2	7	9			60	65	69		
94: Food Services	357	1,184	232	0	1	1			2	2	3		
95: Military Police	102	1,590	305	9	9	15			56	63	65		
96: Military Intelligence	299	1,132	130	7	12	24			69	65	67	*	
97: Bands	131	197	88	7	8	13			80	81	85		
98: Signals Intelligence/ Electronic Warfare Operations	222	723	96	5	8	10			46	59	66		

Table A.4

FY94 E6 One-Year Promotion and Retention Rates by CMF

1994 E6	Total NCOs			One-Year Promotion Rates (percent)					One-Year Retention Rates (percent)				
	Below PSL	At PSL	Above PSL	Below PSL	At PSL	Above PSL			Below PSL	At PSL	Above PSL		
11: Infantry	314	3,559	428	6	2	15	*		62	63	70		
12: Combat Engineering	47	796	128	30	8	20	*		68	71	80		
13: Field Artillery	135	2,319	356	8	3	11	*		70	73	72		*
14: Air Defense Artillery	85	519	56		1	11			69	57	59	*	
18: Special Forces	29	889	369	55	11	19	*		62	78	79		
19: Armor	104	1,287	249	10	5	27	*		69	66	67	*	
23: Air Defense System Maintenance	17	70	8	6	3	13	*		65	59	38	*	*
25: Visual Information	1	5							100	40		*	*
27: Land Combat and Air Defense System Maintenance	43	204	8	5	5	13			65	57	25	*	*
29: Signal Maintenance	81	259	7	1	2			*	12	10		*	*
31: Signal Operations	285	1,795	107		1	7			47	38	72	*	
33: Electronic Warfare/ Intercept Systems Maintenance	30	41	5	13	10		*	*	70	63	60	*	*
35: Electronic Maintenance and Calibration	14	87	4		2	25			29	44	50		
37: Psychological Operations	14	36	4		3			*	79	81	100		
46: Public Affairs	11	6	4			25			18	50	50		
51: General Engineering	15	376	68	7	6	16	*		53	55	65		
54: Chemical	85	712	183	9	3	6	*		73	70	67	*	*

Table A.4—continued

1994 E6	Total NCOs			One-Year Promotion Rates (percent)					One-Year Retention Rates (percent)				
	Below PSL	At PSL	Above PSL	Below PSL	At PSL	Above PSL			Below PSL	At PSL	Above PSL		
55: Ammunition	30	415	52	23	3	17	*		63	70	71		
63: Mechanical Maintenance	269	2,249	522	9	6	16	*		70	65	73	*	
67: Aircraft Maintenance	356	1,367	93	1	2	4			69	66	66	*	
71: Administration	190	973	102	2	7	11			54	47	49	*	
74: Automatic Data Processing	32	247	58	6	3	10	*		47	49	45		*
77: Petroleum and Water	93	260	50	6	10	6		*	76	70	74	*	
79: Recruiting and Reenlistment		56	9										
81: Topographic Engineering	29	68	5	7	4		*	*	55	34	60	*	
88: Transportation	214	1,648	168	2	2	8	*		71	68	64	*	*
91: Medical	338	1,255	178	4	3	13	*		48	53	60		
92: Supply and Services	254	1,780	268	13	7	20	*		64	68	76		
93: Aviation Operations	86	166	49	1	5	8			53	54	59		
94: Food Services	137	641	169						1	1			*
95: Military Police	125	918	135	8	3	13	*		59	60	59		*
96: Military Intelligence	208	578	103	5	10	21			62	61	67	*	
97: Bands	165	398	82	3	4	16			80	85	78		*
98: Signals Intelligence/ Electronic Warfare Operations	158	369	66	4	4	8	*		49	58	62		

Table A.5

FY96 E5 One-Year Promotion and Retention Rates by CMF

	Total NCOs			One-Year Promotion Rates (percent)					One-Year Retention Rates (percent)				
1996 E5	Below PSL	At PSL	Above PSL	Below PSL	At PSL	Above PSL			Below PSL	At PSL	Above PSL		
11: Infantry	1,030	3,911	731	10	15	25			71	69	68	*	*
12: Combat Engineering	152	1,048	159	8	9	21			78	73	75	*	
13: Field Artillery	431	2,542	528	10	15	23			77	74	74	*	*
14: Air Defense Artillery	135	706	127	11	12	14			52	56	49		*
18: Special Forces		1	22		100	68		*		100	82		*
19: Armor	415	1,988	257	10	16	30			69	73	78		
23: Air Defense System Maintenance	35	66	12	3	3			*	17	14	8	*	*
25: Visual Information	19	44	6		5	17			58	50	50	*	
27: Land Combat and Air Defense System Maintenance	11	199	28	36	15	11	*	*	91	73	71	*	*
29: Signal Maintenance													
31: Signal Operations	800	2,846	336	6	10	13			69	71	72		
33: Electronic Warfare/ Intercept Systems Maintenance	31	53	4	10	17	25			65	68	75		
35: Electronic Maintenance and Calibration	142	527	43	8	10	16			70	65	63	*	*
37: Psychological Operations	28	46	27	25	30	26		*	79	74	70	*	*
46: Public Affairs	14	11	5	14	9	20	*		57	73	40		*
51: General Engineering	179	683	22	6	10	18			70	66	59	*	*
54: Chemical	166	934	190	11	10	13	*		75	76	75		*

Table A.5—continued

	Total NCOs			One-Year Promotion Rates (percent)					One-Year Retention Rates (percent)				
1996 E5	Below PSL	At PSL	Above PSL	Below PSL	At PSL	Above PSL			Below PSL	At PSL	Above PSL		
55: Ammunition	105	401	38	8	20	37			77	80	71		*
63: Mechanical Maintenance	1,353	4,664	359	4	11	26			73	73	79	*	
67: Aircraft Maintenance	494	1,370	164	7	13	15			69	70	61		*
71: Administration	561	1,494	240	6	8	10			57	60	59		*
74: Automatic Data Processing	135	317	57	6	8	12			63	59	53	*	*
77: Petroleum and Water	294	1,048	64	2	8	16			81	76	89	*	
79: Recruiting and Reenlistment													
81: Topographic Engineering	27	58	5		12			*	33	43			*
88: Transportation	587	1,930	185	3	8	15			75	76	71		*
91: Medical	575	2,204	212	6	11	13			63	68	62		*
92: Supply and Services	1,002	3,465	643	6	11	11			75	76	77		
93: Aviation Operations	75	214	38	7	8	5		*	69	62	74	*	
94: Food Services													
95: Military Police	271	1,487	181	6	5	9	*		61	63	71		
96: Military Intelligence	280	907	132	9	12	16			58	65	63		*
97: Bands	94	146	78	18	16	15	*	*	82	86	76		*
98: Signals Intelligence/ Electronic Warfare Operations	136	536	85	4	10	15			49	53	67		

Table A.6

FY96 E6 One-Year Promotion and Retention Rates by CMF

	Total NCOs			One-Year Promotion Rates (percent)				One-Year Retention Rates (percent)				
1996 E6	Below PSL	At PSL	Above PSL	Below PSL	At PSL	Above PSL		Below PSL	At PSL	Above PSL		
11: Infantry	484	3,094	322	1	4	19		66	63	66	*	
12: Combat Engineering	49	632	80	2	2	20		80	68	78	*	
13: Field Artillery	251	1,856	245	1	5	25		70	68	70	*	
14: Air Defense Artillery	81	452	98	2	6	15		42	50	46		*
18: Special Forces		887	312		20	23			83	79		*
19: Armor	227	1,165	186		3	29		70	62	67	*	
23: Air Defense System Maintenance	20	40	13					20	13	8	*	*
25: Visual Information	13	28						62	64			*
27: Land Combat and Air Defense System Maintenance	17	126	18		13	28		71	66	72	*	
29: Signal Maintenance												
31: Signal Operations	354	1,249	117	1	5	23		66	64	77	*	
33: Electronic Warfare/Intercept Systems Maintenance	18	36	9					61	58	67	*	
35: Electronic Maintenance and Calibration	52	259	4		3	25		60	62	100		
37: Psychological Operations	23	38	11		8	27		61	58	55	*	*
46: Public Affairs	3	9	2		22		*	100	44	100	*	
51: General Engineering	19	320	47		7	28		58	70	70		
54: Chemical	130	615	101	2	5	16		75	76	83		

Table A.6—continued

1996 E6	Total NCOs			One-Year Promotion Rates (percent)					One-Year Retention Rates (percent)				
	Below PSL	At PSL	Above PSL	Below PSL	At PSL	Above PSL			Below PSL	At PSL	Above PSL		
55: Ammunition	68	386	36	1	7	33			82	74	86	*	
63: Mechanical Maintenance	335	1,736	384	1	12	24			70	71	75		
67: Aircraft Maintenance	315	1,227	142	2	4	25			71	69	73	*	
71: Administration	179	662	173	3	7	15			65	65	73	*	
74: Automatic Data Processing	56	228	46	4	9	9		*	55	59	63		
77: Petroleum and Water	69	304	49	1	12	24			75	77	78		
79: Recruiting and Reenlistment		19	1		16			*		63	100		
81: Topographic Engineering	9	40	10		3	20			33	35	50		
88: Transportation	236	1,336	126	1	5	20			71	73	74		
91: Medical	305	1,039	132	2	5	16			64	70	61		*
92: Supply and Services	510	2,040	288	2	8	25			73	76	78		
93: Aviation Operations	39	109	27	5	8	15			72	65	81	*	
94: Food Services													
95: Military Police	162	893	124	2	2	8	*		67	70	60		*
96: Military Intelligence	221	565	83	3	7	22			62	65	70		
97: Bands	120	358	60	5	6	22			85	87	92		
98: Signals Intelligence/ Electronic Warfare Operations	111	336	63	2	5	8			62	60	65	*	

Appendix B

THE STEADY-STATE INVENTORY PROJECTION MODEL'S MATHEMATICAL FORMULATION

This appendix presents the mathematical formulation of the NCO LD steady-state inventory projection model (IPM). The model projects the steady-state enlisted force within five separate enlisted groupings: the *operations* group and four *specialty* groups. Within each group the inventory is tracked by grade and year of service.[1]

While a traditional inventory projection model might begin with annual accessions and determine how they move through the grades and years of service in a *forward*, supply-push approach, this IPM uses a *backward*, demand-pull approach, with annual accessions being an output of the model rather than an input. Model inputs include the following:

- the number of soldiers in each grade in the operations group and each of the specialty groups;
- for each group, the grade-by-YOS separation/retirement rates from that group, i.e., losses to the Army from that group;
- the YOS distribution of promotions into each grade; and
- for the operations group, the grade-by-YOS flow rates from the operations group to each specialty group.

---

[1]The analyses presented in this report did not take advantage of the specialty group modeling capabilities.

The promotion distributions must sum to 100 percent, indicating the *percentage* of promotions into a grade that must come from each YOS.

Figures B.1 and B.2 illustrate the flows into and out of one node of the operations group and a specialty group, respectively. In the current formulation, the only flows into the operations group come from annual accessions, and the flows into each specialty group come from annual accessions and transfers only from the operations group. Future formulations may require the inclusion of flows among specialty groups.

In Figure B.1 we see that two types of flow can go into an operations node, namely lateral flows and promotion flows. Lateral flows come from the same grade, and promotion flows come from the lower grade. Four types of flow can leave the node: lateral flows, promotions out of the grade, separations from the Army, and flows to the specialty groups.

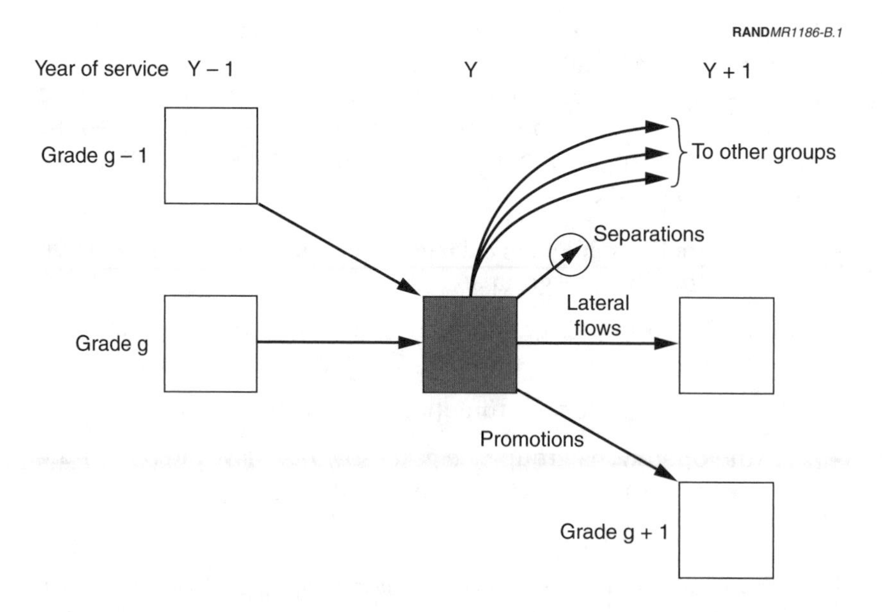

Figure B.1—Flows Associated with an Operations Group Node

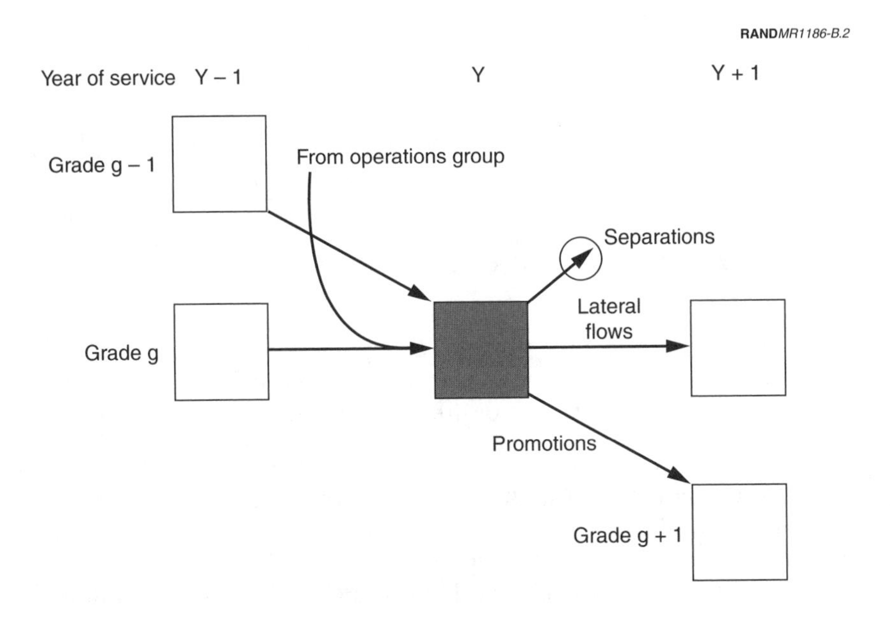

Figure B.2—Flows Associated with a Non-Operations Group Node

In Figure B.2, which represents flows associated with a specialty group's node, we see that three types of flow can come into a node: lateral flows, promotion flows, and flows from the operations group. Three types of flow can leave the node: lateral flows, promotions out of the grade, and separations from the Army. Note that no flows are permitted from a specialty group to another group.

EXAMPLES

The Operations E9 Grade

With this introduction to the flows associated with a node, we now turn to a simple example of the mathematics that underpins the model. Figure B.3 depicts the grade of E9 in the operations group, where we have a total of 3,960 soldiers. Promotions flow into the grade in its first two years, with 25 percent coming in the first year and the remaining 75 percent in the second. Flows out of the node, including both separations/retirements from the Army and flows to

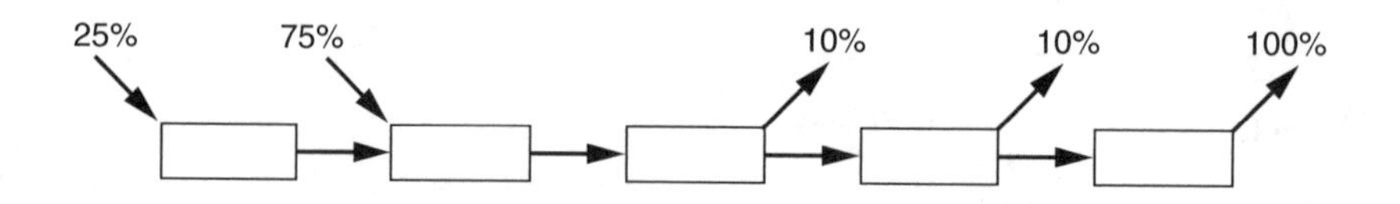

Figure B.3—Flows Associated with Operations Group E9

the specialty groups, take place in the grade's last three years: 10 percent in each of the first two and 100 percent in the last year. There are no promotions out of the grade because E9 is as high as a soldier can go.

With these inputs (3,960 soldiers in E9, the promotion flow distribution into E9 and the *flow out* rates out of the operations group) we can algebraically specify the number of soldiers in each year of service and the actual flows into and out of the grade. Figure B.4 illustrates this, where **P** represents the total number of promotions into E9.

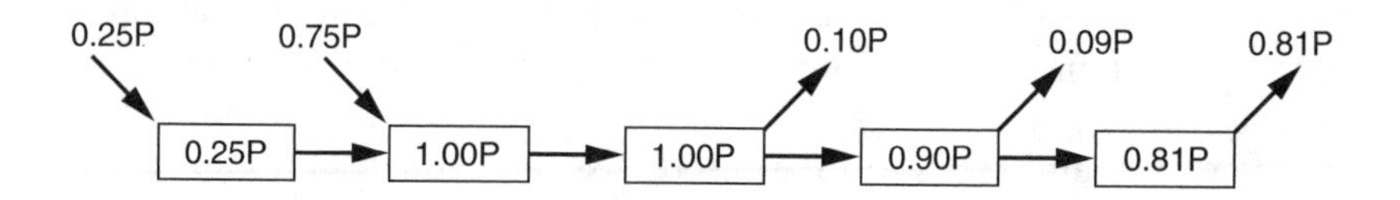

Figure B.4—Algebraic Flows Associated with Operations Group E9

The first year of service has **0.25P** soldiers, which represents 25 percent of the promotion flows into the first E9 year. The second year of service has **1.00P**, coming from the preceding year's lateral and promotion flows (**0.25P** + **0.75P**). The third year also contains **1.00P**, coming entirely from the preceding year's lateral flow. The fourth year reflects the first flows out of the grade, either retirements from the Army or flows to the specialty groups, which happened at the end of the third year. Since 10 percent left the grade in the third year, the fourth year has **0.9P**. The final year also reflects a 10 percent loss to the grade, leaving **0.81P**, which is 90 percent of the preceding year's **0.9P**.

This allows us to write and solve the following linear equation:

$$\mathbf{3{,}960 = 0.25P + 1.00P + 1.00P + 0.90P + 0.81P}.$$

This equation simply reflects that the sum of the soldiers in each of the operations E9 years of service must equal the total number of soldiers in the grade of operations E9. Solving this equation leads to:

$$\mathbf{P = 1{,}000}.$$

With this we can fill in all the flows and states for the grade, as shown in Figure B.5. Note that the number of soldiers flowing into the grade (250 + 750) equals the number of soldiers flowing out of the grade (100 + 90 + 810), which must be the case in a steady-state inventory projection model. Note also that the sum of the soldiers in each year of service (250 + 1,000 + 1,000 + 900 + 810) exactly equals 3,960, the total number of soldiers in the grade.

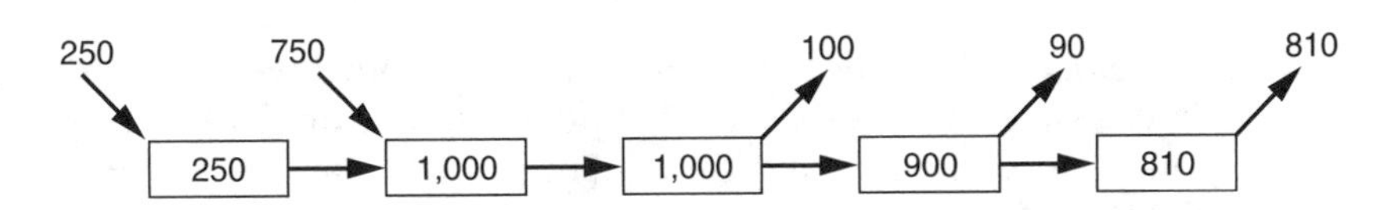

Figure B.5—Operations Group E9 Inventory and Flows

The Operations E8 Grade

The computation begins with the operations group's highest grade, and losses from that grade and the resultant promotions into that grade are determined and distributed over the years of service. The model then moves to the next-lower grade, knowing the promotion flows out of it. This is reflected in Figures B.6 and B.7, where the total number of operations E8s is 6,575. Figure B.6 presents the inputs to the process, including the promotion flows into E9—we've assumed

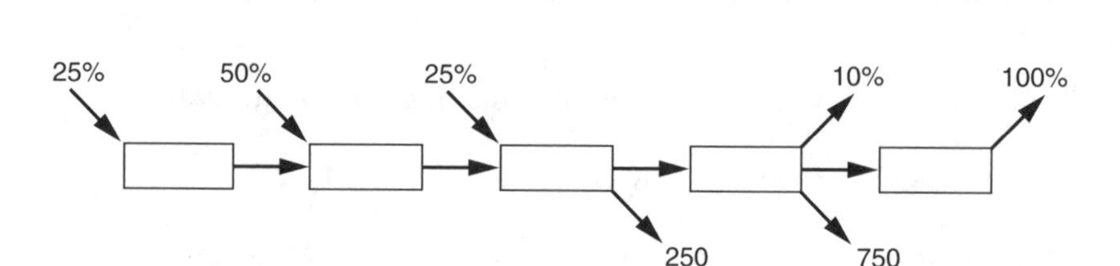

Figure B.6—Operations Group E8 Inputs

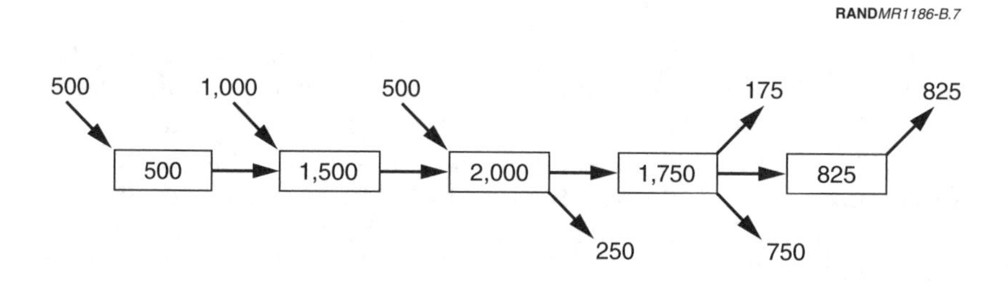

Figure B.7—Operations Group E8 Inventory and Flows

that these take place out of the grade's third and fourth years of service. We've also assumed that losses due to separations or flows to other groups are 10 percent at the end of the fourth year and 100 percent at the end of the fifth year.

We leave the algebraic details as an exercise. The result is shown in Figure B.7. Note that the inputs imply 2,000 promotions into operations E8. Further, the flows into the grade (2,000) exactly equal the flows out of the grade (1,000 separations/flows to other groups and 1,000 promotions to E9). In other words, the methodology determines the number of promotions needed to sustain E8 *and* provide sufficient NCOs to support promotion flows to E9. (Determining the promotions needed to sustain E8 is done without consideration of whether E7 has sufficient inventory to support those promotions.) Also note that the losses from a year of service are determined based on the total number in the year of service and not on the number after promotions are removed (see the fourth year, where the 175 losses reflect 10 percent of 1,750). Finally, the sum of the inventories in each year of service equals the total required inventory for the grade (6,575).

The Specialty Groups

The algebra associated with the specialty groups is slightly more complicated. This complication arises because there are specific numeric flows from the operations group into each specialty group that must be considered. Figure B.2 above illustrates these flows. As the algebra is similar to that for the operations group, we won't illustrate it here.

MEASURES OF EFFECTIVENESS

Several measures are apparent from the example. First, we can easily compute average years in grade for each grade as well as average years of service. We can also determine average years in grade of those who are destined for separation/retirement and average years in grade of those who are destined for promotion. Similarly, we can compute average years of service of those who are destined for separation/retirement and for those who are destined for promotion.

Finally, we can determine the probability of promotion to the grade of E9 (in this example), which is nothing more than the ratio of E9 promotions to E8 promotions—for this example, the probability of promotion is 50 percent.

Probability of promotion differs from selection rate. Probability of promotion is defined to be the probability that a soldier will get promoted to the next-higher grade given that he has *just been promoted* to a grade, e.g., "I just became an E8; what are my chances of making E9?" This measure reflects the soldier's career chances as opposed to his single-year chances. The measure could also be expanded to cover grade aggregates, e.g., "Given that I just made E4, what are my chances of making E9?" The measure is simply the ratio of E9 promotions to E4 promotions.

ALGEBRAIC FORMULATION

In this section we present the algebraic equations that support the inventory projection model. The next section discusses the spreadsheet's computational steps to implement the algebra contained in this section.

Consider the following variables:[2]

$\mathbf{g}$:	The grade identifier
$\mathbf{y,i,j}$:	The YOS identifiers
$\mathbf{k_g}$:	The number of years of service for which grade **g** exists
OPS :	The operations group identifier
$\mathbf{N_{OPS,g}}$:	The inventory in grade **g** of the operations group
$\mathbf{PDIST_{OPS,g,y}}$:	The promotion distribution for promotions into OPS **g**
$\mathbf{CONTINUE_{OPS,g,y}}$:	The continuation rate from year **y** to year **y+1** for OPS **g**—this is simply 1.0 minus the sum of the separation rate and the specialty group transfer rates
$\mathbf{n_{OPS,g,y}}$:	The inventory in OPS **g** year **y**
$\mathbf{p.in_{OPS,g}}$:	The number of promotions into OPS **g** (what the model determines)
$\mathbf{P.OUT_{OPS,g,y}}$:	The number of promotions out of OPS **g** in year **y**

The Operations Group

Note that $\mathbf{p.in_{OPS,g}}$ is the total number of promotions into OPS **g**, while $\mathbf{P.OUT_{OPS,g,y}}$ specifies the promotions out of OPS **g** for each year **y**. We have specified the above definitions using a full-

---

[2]We adopt the convention that independent variables are presented in upper case and dependent variables in lower case.

dimensional notation, e.g., **OPS,g,y**. In the rest of this section we will drop the **OPS,g** and only use the **y** dimension.

The following four equations specify the inventory in each of the years for which OPS **g** exists, beginning with the first year. The first equation simply states that the first soldiers who enter OPS **g** must come from promotions from grade **g – 1**, and these promotions are governed by the **PDIST** promotion distribution. In the second equation, the number in year 2 must come from lateral flows from year 1 ($\mathbf{n_1 * CONTINUE_1 - P.OUT_1}$) or from promotions into OPS **g**'s second year ($\mathbf{PDIST_2 * p.in}$). The third and fourth equations have similar meanings.

$$\mathbf{n_1 = PDIST_1 * p.in}$$
$$\mathbf{n_2 = n_1 * CONTINUE_1 - P.OUT_1 + PDIST_2 * p.in}$$
$$\mathbf{n_3 = n_2 * CONTINUE_2 - P.OUT_2 + PDIST_3 * p.in}$$
$$\ldots$$
$$\mathbf{n_{\mathit{k}} = n_{\mathit{k}-1} * CONTINUE_{\mathit{k}-1} - P.OUT_{\mathit{k}-1} + PDIST_{\mathit{k}} * p.in}\ .$$

Looking at the fourth equation, the year dimension for the last term ($\mathbf{PDIST_{\mathit{k}}}$) is different from the year dimensions for the first two terms ($\mathbf{n_{\mathit{k}-1} * CONTINUE_{\mathit{k}-1} - P.OUT_{\mathit{k}-1}}$). This reflects the fact that flows into a node take place at the beginning of the year (or come from the *end* of the previous year) and flows out of a node take place at the end of the year (or the beginning of the *next* year).

Clearly, the sum of the inventory in each node must equal the total inventory for the grade:

$$\mathbf{N} = \sum_i \mathbf{n}_i\ .$$

However, the $\mathbf{n}_i$'s in this equation are specified in terms of previous years' **n**'s. The next step therefore is to expand the **n**'s into expressions with only independent variables. The following equations show this expansion. Those equations are the expansions of the four equations above. We've expanded the first three to show the summation/product pattern that is reflected in the fourth.

$$\begin{aligned}
\mathbf{n}_1 &= \mathbf{PDIST}_1 * \mathbf{p.in} \\
\mathbf{n}_2 &= \mathbf{PDIST}_1 * \mathbf{p.in} * \mathbf{CONTINUE}_1 - \mathbf{P.OUT}_1 + \mathbf{PDIST}_2 * \mathbf{p.in} \\
&= \mathbf{p.in} * \left(\mathbf{PDIST}_1 * \mathbf{CONTINUE}_1 + \mathbf{PDIST}_2\right) - \mathbf{P.OUT}_1 \\
\mathbf{n}_3 &= \left\{\mathbf{p.in} * \left(\mathbf{PDIST}_1 * \mathbf{CONTINUE}_1 + \mathbf{PDIST}_2\right) - \mathbf{P.OUT}_1\right\} \\
&\quad * \mathbf{CONTINUE}_2 - \mathbf{P.OUT}_2 + \mathbf{PDIST}_3 * \mathbf{p.in} \\
&= \mathbf{p.in} * \left(\mathbf{PDIST}_1 * \mathbf{CONTINUE}_1 * \mathbf{CONTINUE}_2 + \mathbf{PDIST}_2 * \mathbf{CONTINUE}_2 + \mathbf{PDIST}_3\right) \\
&\quad - \left(\mathbf{P.OUT}_1 * \mathbf{CONTINUE}_2 + \mathbf{P.OUT}_2\right) \\
&\quad \ldots \\
\mathbf{n}_k &= \mathbf{n}_{k-1} * \mathbf{CONTINUE}_{k-1} - \mathbf{P.OUT}_{k-1} + \mathbf{PDIST}_k * \mathbf{p.in} \\
&= \mathbf{p.in} * \sum_{i=1}^{k} \mathbf{PDIST}_i \prod_{j=i}^{k-1} \mathbf{CONTINUE}_j - \sum_{i=1}^{k-1} \mathbf{P.OUT}_i * \prod_{j=i+1}^{k-1} \mathbf{CONTINUE}_j .
\end{aligned}$$

The fourth equation has two product operators. The first product operator ranges from **i** to **k – 1**. In one instance, when **j=k**, this product is assumed to be 1.0. Similarly, the limits on the second term's product operator are **j + 1** and **k – 1**. In one instance, when **j = k – 1**, this product is assumed to be 1.0.

The fourth equation has two terms that provide some interesting insight into the model's mathematics. This equation expresses the number of soldiers in OPS **g** in year **k**. The first term reflects the continuation pattern for those soldiers promoted into the grade without taking into consideration that some of them will eventually be promoted out. The second term reflects what we've chosen to call the *promotion residue*, the continuation pattern of those promoted out of the grade *had they not been promoted*, i.e., had they remained in OPS **g**. In other words, the model determines promotion requirements *into* grade **g** to allow not only for fulfilling the grade's inventory, but also for providing for promotions *out of* the grade. It essentially *creates* inventory in grade **g** that would not otherwise exist and will have to be removed in a later phase of the computations. This residual inventory is what gets promoted to grade **g + 1**, hence the term *promotion residue*. We will deal with this in subsequent discussions. For E9s, there are no promotions out of grade, and this second term does not appear.[3]

---

[3]The product terms Π**CONTINUE** are referred to as the retention rate template in the spreadsheet discussion below.

The final step in the process is to add all the $\mathbf{n}_i$'s together. This leads to the following general equation:

$$\mathbf{N} = \mathbf{p.in} * \mathbf{m} - \mathbf{p.residue}$$

$$\mathbf{p.in} = \frac{\mathbf{N} + \mathbf{p.residue}}{\mathbf{m}},$$

where

$$\mathbf{m} = \sum_{y=1}^{k} \sum_{i=1}^{y} \mathbf{PDIST}_i \prod_{j=i}^{k-1} \mathbf{CONTINUE}_j$$

$$\mathbf{p.residue} = \sum_{y=1}^{k-1} \sum_{i=1}^{y} \mathrm{P.OUT}_i \prod_{j=i+1}^{k-1} \mathrm{CONTINUE}_j.$$

The Specialty Groups

The above solution for **p.in** applies to the operations group. For each specialty group, this equation takes a slightly different form, to account for the number of soldiers who transfer in from the operations group. The equation's general form is

$$\mathbf{N} = \mathbf{p.in} * \mathbf{m} - \mathbf{p.residue} + \mathbf{ops.residue}$$

$$\mathbf{p.in} = \frac{\mathbf{N} + \mathbf{p.residue} - \mathbf{ops.residue}}{\mathbf{m}}.$$

The **ops.residue** term is similar to the **p.residue** term in that continuation rates are applied to the OPS flows into the specialty group in the same manner that they were applied to the promotion flows out. While the grade **g + 1** promotion residue is added to the numerator to reflect that promotions out of grade **g** must also be accounted for when determining the number of promotions into grade **g**, the **ops.residue** term is subtracted to reflect that the promotions into grade **g** must not account for those soldiers coming from OPS **g.**

SPREADSHEET DETAILS

This section discusses the model spreadsheet in some detail. It shows example worksheet arrays and discusses the math that underlies computational (as opposed to input) arrays. We'll first discuss the operations group's spreadsheet. Then we'll turn to one of the specialty groups.

Operations Group

The array in Table B.1 shows the E4 inputs for the operations group.[4] This worksheet array is representative of all worksheet arrays for the operations group spreadsheet.

The **E4 inventory** is specified as 128,000. Next to the first column, YOS, are seven additional columns. The first guides **promotions into E4**, showing that 75 percent come into the fourth year of service, and the remaining 25 percent come into the fifth year of service. This means that the promotions show up in the grade of E4 at the *beginning* of the fourth and fifth years, and that those same promotions leave E1–3 at the *end* of the third and fourth years. There are no other flows into the OPS E4 grade.

The next four columns specify **the percentage of flows to specialty groups** from the operations group. Flows to specialty group 1 leave the operations group at the end of the fourth, fifth, and sixth years, showing up in that group at the beginning of the fifth, sixth, and seventh years. Ten percent of the OPS E4s in the fourth year go SPEC 1, 2 percent in the fifth year, and 1 percent in the sixth year. Similar percentages are provided for the other specialty groups.

The next column specifies **the loss rates out of the OPS E4 grade**. These inputs reflect losses to the Army and not flows to other groups. For years 4 through 19, 1 percent of the OPS E4 soldiers with the indicated year leave the Army at the end of that year. At the end of the 20th year, all OPS E4 soldiers with 20 years of service must leave the Army.

There is another input associated with OPS E4s. It specifies the number of **promotions out of OPS E4** that must be made to support the inventory needs of OPS E5. This input is not specified by the user. Rather, it is determined as part of the computation process for

---

[4]All inputs in this appendix are contrived. They do not reflect the actual rates associated with promotion, separation/retirement, or intergroup transfer.

Table B.1

Inputs for the Operations Group

E4 Inventory = 128,000		Flows to Other Groups Out of this YOS					
YOS	Promotion Distribution into This YOS	To Group 1	To Group 2	To Group 3	To Group 4	Loss Rate from This YOS	Promotions Out of E4
1							
2							
3							
4	**75%**	**10%**	**5%**	**5%**	**2%**	**1%**	
5	**25%**	**2%**	**1%**	**1%**		**1%**	882
6		**1%**	**1%**			**1%**	882
7						**1%**	1,765
8						**1%**	1,765
9						**1%**	1,765
10						**1%**	882
11						**1%**	882
12						**1%**	
13						**1%**	
14						**1%**	
15						**1%**	
16						**1%**	
17						**1%**	
18						**1%**	
19						**1%**	
20						**100%**	

OPS E5, which took place before OPS E4 computations. This is presented as the last column and is shown in normal as opposed to boldface type. This column shows that 882 E4s will be promoted to E5 during their fifth year of service, i.e., they will be E4s until the end of the fifth year and E5s at the beginning of their sixth year. The same number of promotions will occur in the sixth, tenth, and eleventh years. For years seven, eight, and nine, there will be 1,765 promotions out of these years into OPS E5.[5]

---

[5]This array, an input to this computation from the E5 computation, reflects the OPS E5 promotion distribution, calling for 10 percent of promotions to OPS E5 into the sixth, seventh, eleventh, and twelfth years, with 20 percent in the eighth, ninth, and tenth years.

A final input, which governs **promotion eligibility**, is available but not illustrated here—it is discussed when describing model calculations below (Table B.2). This input specifies when OPS E4s are eligible for promotion to E5. The input specifies that an E4 must serve at least one year in E4 before he can be considered for promotion to E5. This is reflected in the table by having the first years in grade shown in italics. Those soldiers are not eligible for promotion.

We are now ready to turn to the calculations performed by the OPS E4 worksheet. The worksheet segment below shows the final results of this process. The array shows the inventory in terms of year of service and *entry* year of service. The full array is 30-by-30, but we have shown only the relevant years of service and the early entry years.

Table B.2

Promotion Eligibility

	Entry Year of Service						
YOS	1	2	3	4	5	6	Total
1							
2							
3							
4				*13,815*			13,815
5				10,638	*4,605*		15,242
6				9,490	4,108		13,598
7				8,589	3,718		12,308
8				7,272	3,148		10,420
9				5,968	2,583		8,551
10				4,676	2,024		6,701
11				4,014	1,738		5,752
12				3,358	1,454		4,812
13				3,324	1,439		4,764
14				3,291	1,425		4,716
15				3,258	1,411		4,669
16				3,226	1,396		4,622
17				3,193	1,382		4,576
18				3,162	1,369		4,530
19				3,130	1,355		4,485
20				3,099	1,341		4,440
21							
Sum				93,503	34,497		128,000

The two entry years of service arise because the promotion distribution inputs specify promotions to take place into years four and five only, 75 percent into year 4 and 25 percent into year 5. Not surprisingly, the actual number of promotions into year 4 (13,815) is three times the number into year 5 (4,605).

The inventory stops with year 20 because the inputs specify that all OPS E4s must leave at the end of the 20th year. The dark-shaded area reflects the fact that no OPS E4s can mathematically exist in those areas, i.e., no E4s can have a year of service that is less than the entry year of service. The light-shaded areas reflect the fact that the inputs preclude the existence of OPS E4s in all entry years of service except years four and five. The inputs also preclude OPS E4s with 21 years of service.

We will now describe the calculation steps that lead to this result. The worksheet segment in Table B.3 shows the OPS E4 continuation rates. As the column heading implies, these rates reflect the percentage of OPS E4s who remain in the operations group from one year to

Table B.3

OPS E4 Continuation Rates

YOS	1.0 – Sum of All Flows Out of YOS		YOS	1.0 – Sum of All Flows Out of YOS
1			16	99%
2			17	99%
3			18	99%
4	77%		19	99%
5	95%		20	
6	97%		21	
7	99%		22	
8	99%		23	
9	99%		24	
10	99%		25	
11	99%		26	
12	99%		27	
13	99%		28	
14	99%		29	
15	99%		30	

the next. Some of those who move will actually be promoted to E5, but the continuation rates reflect all those who remain in the operations group, *including* those who will be promoted. We deal with promotions separately. Thus, these continuation rates reflect, by implication, soldiers moving out of the operations group either by leaving the Army or by moving to one of the four specialty groups.

The model takes these YOS-based continuation rates and computes a YOS-by-entry-YOS set of cumulative continuation rates, which we have chosen to call the *retention rate template*. This template appears in the worksheet segment shown in Table B.4. It is a 30-by-30 array, and we show only the relevant segment here. The template indicates the fraction of OPS E4s with a given entry YOS that will remain in OPS E4 in the given YOS, which equates to the product

Table B.4

Retention Rate Template

YOS	4	5	6	7	8	9	10	11	12
4	1.00								
5	0.77	1.00							
6	0.73	0.95	*1.00*						
7	0.71	0.92	*0.97*	*1.00*					
8	0.70	0.91	*0.96*	*0.99*	*1.0*				
9	0.70	0.90	*0.95*	*0.98*	*0.99*	*1.0*			
10	0.69	0.89	*0.94*	*0.97*	*0.98*	*0.99*	*1.00*		
11	0.68	0.89	*0.93*	*0.96*	*0.97*	*0.98*	*0.99*	*1.00*	
12	0.67	0.88	*0.92*	*0.95*	*0.96*	*0.97*	*0.98*	*0.99*	*1.00*
13	0.67	0.87	*0.91*	*0.94*	*0.95*	*0.96*	*0.97*	*0.98*	*0.99*
14	0.66	0.86	*0.90*	*0.93*	*0.94*	*0.95*	*0.96*	*0.97*	*0.98*
15	0.65	0.85	*0.90*	*0.92*	*0.93*	*0.94*	*0.95*	*0.96*	*0.97*
16	0.65	0.84	*0.89*	*0.91*	*0.92*	*0.93*	*0.94*	*0.95*	*0.96*
17	0.64	0.83	*0.88*	*0.90*	*0.91*	*0.92*	*0.93*	*0.94*	*0.95*
18	0.64	0.83	*0.87*	*0.90*	*0.90*	*0.91*	*0.92*	*0.93*	*0.94*
19	0.63	0.82	*0.86*	*0.89*	*0.90*	*0.90*	*0.91*	*0.92*	*0.93*
20	0.62	0.81	*0.85*	*0.88*	*0.89*	*0.90*	*0.90*	*0.91*	*0.92*
21									
	11.81	14.05	**SUM**						
	0.75	0.25	**P.in.trans**						
	8.86	3.51	**SUM * P.in.trans**						

terms **ΠCONTINUE** in the algebraic formulation. It is the product of successive continuation rates, beginning with the entry YOS.[6]

The first row at the bottom of the worksheet segment (**SUM**) is simply the sum of all the cumulative continuation rates for the entry YOS. The second row on the bottom (**P.in.trans**) is the promotion distribution into OPS E4, and the final bottom row (**SUM * P.in.trans**) is the product of the promotion distribution and the sum—these are the **ΣPDIST * ΠCONTINUE** terms in the algebraic formulation.

The model must next compute the E4 inventory implications associated with the promotions *out of OPS E4* assuming that they would have stayed in E4—the **ΣP.OUT * ΠCONTINUE** terms in the algebraic formulation. The worksheet segment illustrated in Table B.5, which

Table B.5

E5 Promotion "Residue" in E5 Entry YOS Form

YOS	6	7	8	9	10	11	12	Total
6	882							882
7	856	882						1,738
8	847	874	1,765					3,485
9	839	865	1,747	1,765				5,215
10	830	856	1,730	1,747	1,765			6,928
11	822	848	1,712	1,730	1,747	882		7,741
12	814	839	1,695	1,712	1,730	874	882	8,546
13	806	831	1,678	1,695	1,712	865	874	8,460
14	798	822	1,661	1,678	1,695	856	865	8,376
15	790	814	1,645	1,661	1,678	848	856	8,292
16	782	806	1,628	1,645	1,661	839	848	8,209
17	774	798	1,612	1,628	1,645	831	839	8,127
18	766	790	1,596	1,612	1,628	822	831	8,046
19	759	782	1,580	1,596	1,612	814	822	7,965
20	751	774	1,564	1,580	1,596	806	814	7,886
21								
Total	12,116	11,581	21,614	20,049	18,469	8,437	7,631	99,897

[6]For inventory projection purposes, the only relevant entry years are four and five. But to determine the implications of promotions out of OPS E4, we also need to know the cumulative continuation rates for years six through twelve. This is why they are presented in italics. The sums at the worksheet segment's bottom span only entry years four and five.

shows the E5 promotion *residue* in E5 entry year format, reflects this. It shows the inventory implications of the E5 promotions had they not been promoted to E5, except that those implications are shown in E5 entry year format. Later, we will have to convert these to E4 entry year format.

Note that the promotions out of OPS E4 are shown in their appropriate spot *were they E5s*, i.e., 882 promotions into OPS E5 years 6, 7, 11, and 12, and 1,765 promotions into years 8, 9 and 10. This implies that they are OPS E4 until the end of years 5–11. The worksheet segment is computed by simply applying the cumulative continuation rates in the worksheet segment above. Had these soldiers not been promoted to OPS E5, they would have added 99,897 more soldiers to the OPS E4 inventory.

With this computation we can now determine the number of promotions needed into OPS E4. It is given by the following equation:

$$\mathbf{p_{OPS.E4}} = \frac{\mathbf{N_{OPS.E4} + E5.residue_{OPS.E4}}}{\mathbf{\sum SUM * P.in.trans}},$$

where

$p_{OPS.E4}$ The number of promotions into OPS E4, the solution to the equation.

$N_{OPS.E4}$ The OPS E4 inventory (128,000).

$E5.residue_{OPS.E4}$ The promotion residue for those soldiers promoted to OPS E5, i.e., their OPS E4 inventory implications had they not been promoted to OPS E5 (99,897 in our example).

SUM*P.in.trans The sums from the bottom of the retention template worksheet segment (8.86 + 3.51 in our example).

This leads to the following solution:

$$\mathbf{p_{OPS.E4}} = \frac{\mathbf{128{,}000 + 99{,}897}}{\mathbf{8.86 + 3.51}} = \mathbf{18{,}420}.$$

Applying the promotion distribution and employing the retention template leads to the worksheet segment shown in Table B.6, which shows the inventory projection of those promotions *assuming that no promotions to OPS E5 occur*. Note that the promotions associated with entry year 4, 13,815, are 75 percent of $\mathbf{p}_{OPS.E4}$ and that those associated with entry year 5, 4,605, are 25 percent of $\mathbf{p}_{OPS.E4}$. Further, note that the total inventory, shown in the bottom right corner of the worksheet segment (227,897), is the sum of the OPS E4 inventory specified in the inputs (128,000) and the total E5 promotion residue inventory (99,897).

The next step in the computation process is to remove the promotions to OPS E5 from the above OPS E4 inventory. This requires that we transform the E5 promotion residue from E5 entry year format to E4 entry year format. This is done by proportionally distributing the E5 promotions within a YOS over the OPS E4s who are eligible for

Table B.6

E4 Inventory Before Taking E5 Promotions Into Consideration

YOS	4	5	Total
4	13,815		13,815
5	10,638	4,605	15,242
6	10,106	4,375	14,480
7	9,802	4,243	14,046
8	9,704	4,201	13,905
9	9,607	4,159	13,766
10	9,511	4,117	13,629
11	9,416	4,076	13,492
12	9,322	4,036	13,358
13	9,229	3,995	13,224
14	9,137	3,955	13,092
15	9,045	3,916	12,961
16	8,955	3,877	12,831
17	8,865	3,838	12,703
18	8,777	3,799	12,576
19	8,689	3,761	12,450
20	8,602	3,724	12,326
21			
Total	163,220	64,677	227,897

Table B.7

Transforming the E5 Promotion Residue to E4 Entry Year Format

	E5 Promotion Eligibles			E5 Promotion Residue		
YOS	4	5	Total	Total	4	5
1						
2						
3						
4						
5	10,638		10,638			
6	10,106	4,375	14,480	882	616	267
7	9,802	4,243	14,046	1,738	1,213	525
8	9,704	4,201	13,905	3,485	2,432	1,053
9	9,607	4,159	13,766	5,215	3,640	1,576
10	9,511	4,117	13,629	6,928	4,835	2,093
11	9,416	4,076	13,492	7,741	5,402	2,339
12	9,322	4,036	13,358	8,546	5,964	2,582
13	9,229	3,995	13,224	8,460	5,904	2,556
14	9,137	3,955	13,092	8,376	5,845	2,530
15	9,045	3,916	12,961	8,292	5,787	2,505
16	8,955	3,877	12,831	8,209	5,729	2,480
17	8,865	3,838	12,703	8,127	5,672	2,455
18	8,777	3,799	12,576	8,046	5,615	2,431
19	8,689	3,761	12,450	7,965	5,559	2,406
20	8,602	3,724	12,326	7,886	5,503	2,382
21						
SUM	149,405	60,072	209,477	99,897	69,717	30,180

promotion from that YOS. The array in Table B.7, which is a composite of several OPS E4 worksheet arrays, shows this computation. The first two columns (after the YOS column) show the number of eligibles in YOS by entry-year terms. Note that soldiers in their first year as an OPS E4 are not eligible for promotion to E5, reflected in the second and third worksheet columns—the blank entries in the fourth and fifth years. This eligibility is governed by an input array.

Next, by using the YOS totals for the E5 promotion residue, we distribute the residue over the E4 entry years, using a proportional distribution methodology based on the three E5 promotion eligibles columns, e.g.,

$$616 = 882 * \frac{10{,}106}{14{,}480}$$

$$267 = 882 * \frac{4{,}376}{14{,}480},$$

resulting in the final two columns.

The final step removes the E5 promotion residue from the E4 inventory above, yielding the worksheet segment shown in Table B.8. Note that the total OPS E4 inventory is 128,000, consistent with the input requirement. Note also that 75 percent of the promotions come from entry year 4 (13,815) and 25 percent come from entry year 5 (4,605).

Table B.8

Total Operations E4 Inventory

YOS	4	5	Total
4	13,815		13,815
5	10,638	4,605	15,242
6	9,490	4,108	13,598
7	8,589	3,718	12,308
8	7,272	3,148	10,420
9	5,968	2,583	8,551
10	4,676	2,024	6,701
11	4,014	1,738	5,752
12	3,358	1,454	4,812
13	3,324	1,439	4,764
14	3,291	1,425	4,716
15	3,258	1,411	4,669
16	3,226	1,396	4,622
17	3,193	1,382	4,576
18	3,162	1,369	4,530
19	3,130	1,355	4,485
20	3,099	1,341	4,440
21			
SUM	93,503	34,497	128,000

Specialty Groups

The specialty group spreadsheets are similar to those for the operations group. There are two major differences: (1) Where the operations group spreadsheet must indicate the flow *rates* for transfers from the operations group to the specialty groups, each specialty group must be given the actual number of soldiers that are flowing to the group. (2) While both the operations group and specialty groups must determine the promotion residue associated with promotions out of the grade in order to ensure that those soldiers are reflected when determining the number of promotions into the grade (**p.residue** in the algebraic formulation), the specialty groups must also determine the operations residue (**ops.residue** in the algebraic formulation) so that these soldiers can be reflected when determining the promotions into the specialty grade. The spreadsheets below reflect this.

The worksheet segment shown in Table B.9, for SPEC1 E4, presents the inputs to Specialty Group 1. The important item in this worksheet segment is the column headed "Flows from the OPS Group (from YOS)."

The worksheet segment shows that 1,381 OPS E4s are transferring into SPEC1 E4 from year 4 (they are OPS E4 in year 4 and SPEC1 E4 in year 5). The related flow percentage specified in the OPS E4 inputs is 10 percent, and the 1,381 is 10 percent of the year 4 OPS E4 inventory in the above worksheet segment (13,815). For years 5 and 6, the flow percentages are 2 and 1 percent, and the year 5 and 6 flows (305 and 136) are also consistent with the corresponding OPS E4 inventories and flow rates.

The worksheet segment in Table B.10 shows the inventory implications of the flows from the operations group. It simply applies the SPEC1 retention template to these flows. Note that the entry years of service are 5 through 7, while the above worksheet segment associates those flows with years 4 through 6. This is consistent with the soldiers leaving OPS E4 at the end of years 4 through 6, and entering SPEC1 E4 at the beginning of years 5 through 7. Note also that the entry year inventories (1,381, 305, and 136) are identical to the flows

Table B.9

Inputs for Specialty Group 1

SPEC E4 Inventory = 6,000					
YOS	Promotion Distribution Into This YOS	Loss Rate from This YOS	Flows from the OPS Group (from YOS)	1.0 – Sum of All Flows Out of YOS	Promotions Out of E4
1				100%	
2				100%	
3				100%	
4	25%	50%	1,381	50%	
5	75%	50%	305	50%	10
6		50%	136	50%	10
7		50%		50%	21
8		50%		50%	21
9		10%		90%	21
10		5%		95%	10
11		5%		95%	
12		5%		95%	
13		5%		95%	
14		5%		95%	
15		5%		95%	
16		5%		95%	
17		5%		95%	
18		5%		95%	
19		5%		95%	
20		100%			

shown in the input worksheet segment. The total SPEC1 E4 inventory generated by these flows (4,716) is the algebraic formulation's **ops.residue**.

From this point computations are essentially analogous to those previously discussed in our illustration of the OPS group example.

Table B.10

Inventory Implications of Flows from OPS Group

YOS	5	6	7	Total
1				
2				
3				
4				
5	1,381			1,381
6	691	305		996
7	345	152	136	634
8	173	76	68	317
9	86	38	34	158
10	78	34	31	143
11	74	33	29	135
12	70	31	28	129
13	67	29	26	122
14	63	28	25	116
15	60	27	24	110
16	57	25	22	105
17	54	24	21	100
18	52	23	20	95
19	49	22	19	90
20	47	21	18	85
21				
SUM	3,347	867	502	4,716

Appendix C

AGGREGATE FORCE AND CMF 67 (AIRCRAFT MAINTENANCE) INVENTORY PROJECTION MODEL INPUT PARAMETERS AND OUTPUTS

AGGREGATE FORCE IPM MODEL

Aggregate Force Model Inputs

The IPM requires three types of input: the inventory in each grade, the separation/retirement rates by year of service out of each grade, and the year-of-service distribution of promotions into each grade. All inputs for the aggregate force are derived from MOSLS outputs in the final three years of its inventory projection. We use the averages for 1 October in each of these years. We chose the last three years of the MOSLS projection because those years are the most stable from the annual accession and postdrawdown perspectives. The specific MOSLS run from which our data derive is MOSLS run M9712, which used data current as of December 1997.

Inventory

Table C.1 shows the inventory by grade for the aggregate force. The inventories shown in this table are normalized to total 410,700. The specific MOSLS beginning FY strengths do not add to 410,700 due to enlisteds in the TTHS pipeline.

Table C.1

Aggregate Enlisted Inventory by Grade

Total	E1–3	E4	E5	E6	E7	E8	E9
410,700	113,432	115,057	74,108	56,087	37,670	11,135	3,210

Loss Rates

Table C.2 shows the loss rates by grade and year of service for the aggregate force. The loss rates are the percentage of NCOs in the specified grade/YOS state at the beginning of the year who leave the

Table C.2

Aggregate Loss Rates Out of Grade and YOS (percent)

YOS	E1–3	E4	E5	E6	E7	E8	E9
1	26.0	2.4					
2	14.7	6.3	17.2				
3	46.8	35.7	42.6	3.4			
4	87.8	31.4	35.5	1.7			
5	84.5	22.2	18.0	10.6			
6	100.0	27.4	16.0	10.9			
7		29.6	15.2	10.8			
8		33.2	16.3	11.0			
9		60.7	13.8	9.1	3.7		
10		100.0	14.6	8.4	10.6		
11			14.7	8.0	8.5		
12			14.2	6.7	4.9		
13			15.8	5.8	4.9		
14			16.9	4.2	3.5		
15			100.0	3.5	2.3		
16				2.8	1.6	1.9	
17				2.9	1.4	0.7	
18				3.2	1.2	1.9	
19				1.0	0.6	0.3	
20				100.0	4.9	2.3	0.8
21					95.0	46.2	9.3
22					100.0	25.8	13.9
23						29.9	15.2
24						100.0	12.2
25							18.1
26							16.2
27							24.7
28							22.2
29							24.2
30							100.0

NOTE: Grade E7, year 20, has a surprisingly low loss rate of 4.9 percent, surprising because year 20 is the first retirement year. Year 21 follows with a 95 percent loss rate. Even though year 20 is the first retirement year, year 20 also has a substantial number of promotions to E8. Tables C.4 and C.5 will show that 392 of 4,028 E7s are promoted to E8 out of year 20. E7s in year 20 hang around to see if they make it. Hence the low year-20 E7 loss rate.

Army during that year, i.e., the losses *out of* the grade and YOS. For purposes of determining force characteristics, the model assumes that all losses from a grade/YOS state take place on the last day of the year (*out of* the grade and YOS).

Promotion Distributions

Tables C.3a and C.3b show the input and *adjusted* promotion distributions by grade and YOS. Promotions *out of* a grade take place on the last day of the year, and promotions *into* a grade take place on the first day of the year. This means that an NCO being promoted out of grade **G** spends a full year in grade **G** before being promoted to grade **G + 1**, and that promotions *into* grade **G + 1** take place at the beginning of the next year. This is consistent with the way losses are treated in the sense that losses from a grade/YOS state take place at the end of the year.

The table shows 100 percent of *promotions* into grades E1–3 taking place in the first year of service. These are annual accessions. The model user could just as easily have specified that annual accessions are distributed over two or more years of service, reflecting prior-service accessions. In our analysis we have chosen not to treat prior service accessions.

Promotions into E4 begin into the second year of service (60 percent) and therefore out of the first year of service in E1–3. Note that the sum of all promotion distribution entries for a grade must be 100 percent.

The adjusted promotion distributions in Table C.3b are similar to those in Table C.3a. They differ in that the high-end tails of the distributions in Table C.3a have been *clipped off* and redistributed over the shaded segments of Table C.3b. This *clipping* was done to provide more reasonable distributions for the steady-state model. Indeed, because this is a steady-state model working from the highest to the lowest grade, it determines the promotions into a grade/YOS state before it knows if there is sufficient inventory in the lower grade's associated state. We found that applying the unadjusted distributions caused infeasibilities, i.e., the need for promotions into a grade/YOS state where there was insufficient inventory in

Table C.3a

Aggregate Promotion Distributions into Grade and YOS (percent)

YOS	E1–3	E4	E5	E6	E7	E8	E9
1	100.0						
2		60.0					
3		40.0	18.0				
4			65.3	0.3			
5			4.0	1.7			
6			3.0	7.5			
7			3.0	12.4			
8			2.6	13.4	0.1		
9			2.6	14.2	0.9		
10			1.3	12.8	4.4		
11			0.3	11.2	6.5		
12				8.4	8.8		
13				6.6	9.9	0.0	
14				4.4	11.2	0.5	
15				3.6	10.8	3.9	
16				3.5	10.2	7.7	
17					9.3	12.1	
18					8.5	14.3	1.7
19					8.2	17.4	8.6
20					6.9	16.9	14.4
21					4.4	14.2	17.8
22						9.1	20.0
23						3.9	14.7
24							12.6
25							10.2
26							
27							
28							
29							
30							
	100	100	100	100	100	100	100

the lower grade's associated state. Eliminating these infeasibilities necessitated the high-end adjustments to the promotion distributions.

Table C.3b

Adjusted Aggregate Promotion Distributions into Grade and YOS (percent)

YOS	E1–3	E4	E5	E6	E7	E8	E9
1	100.0						
2		60.0					
3		40.0	18.3				
4			66.4	0.5			
5			4.1	2.7			
6			3.1	11.9			
7			3.0	19.5			
8			2.6	21.1	0.1		
9			2.6	22.4	1.2		
10				12.8	5.8		
11				9.0	8.6		
12					11.6		
13					13.1	0.0	
14					11.2	0.7	
15					10.8	5.0	
16					10.2	9.9	
17					9.3	15.6	
18					8.5	14.3	1.7
19					8.2	17.4	8.6
20					1.4	16.9	14.4
21						14.2	17.8
22						6.0	20.0
23							22.8
24							14.7
25							
26							
27							
28							
29							
30							
	100	100	100	100	100	100	100

Aggregate Force Model Outputs

We present selected outputs from the inventory projection model. Tables C.4, C.5, and C.6 show the aggregate inventory, promotions, and losses by grade and YOS.

Table C.4

Aggregate Inventory by Grade and YOS

YOS	E1–3	E4	E5	E6	E7	E8	E9	Total
1	84,494							84,494
2	27,871	34,669						62,541
3	664	50,265	5,319					56,248
4	353	13,005	22,291	62				35,711
5	43	7,743	15,247	365				23,398
6	7	5,140	12,058	1,662				18,866
7		2,859	8,801	3,677				15,336
8		1,256	5,843	5,649	9			12,757
9		87	3,118	7,471	96			10,771
10		34	1,245	7,826	500			9,606
11			50	7,586	1,047			8,683
12			43	6,161	1,773			7,976
13			37	4,834	2,602	0		7,473
14			31	3,768	3,240	20		7,059
15			26	2,852	3,744	158		6,780
16				2,038	4,096	432		6,567
17				1,330	4,251	854		6,435
18				698	4,391	1,232	10	6,332
19				104	4,430	1,639	61	6,235
20				5	4,038	2,015	147	6,205
21					3,447	2,255	252	5,954
22					7	1,259	348	1,613
23						798	436	1,233
24						472	457	929
25							401	401
26							329	329
27							276	276
28							208	208
29							162	162
30							123	123
Total	113,432	115,057	74,108	56,087	37,670	11,135	3,210	410,700

Conservation of Flow Relationships

These three tables can be used to illustrate the conservation-of-flow principle that must apply to all Markovian processes. Stated simply, the sum of the flows into a state must equal the sum of the flows out of a state. Alternatively, with the three tables we can demonstrate how NCOs move from one YOS to the next and from one grade to the next.

Table C.5

Aggregate Promotions into Grade and YOS

YOS	E1–3	E4	E5	E6	E7	E8	E9	Total
1	84,494							84,494
2		34,669						34,669
3		23,113	5,319					28,432
4			19,299	62				19,361
5			1,183	304				1,487
6			887	1,335				2,222
7			872	2,197				3,069
8			756	2,378	9			3,143
9			753	2,527	87			3,366
10				1,443	408			1,851
11				1,013	600			1,613
12					814			814
13					917	0		917
14					786	19		805
15					757	139		896
16					714	274		988
17					652	431		1,082
18					594	394	10	998
19					572	481	51	1,104
20					98	466	86	650
21						392	106	498
22						166	119	285
23							136	136
24							88	88
25								
26								
27								
28								
29								
30								
Total	84,494	57,782	29,068	11,260	7,008	2,762	596	192,970

Focusing on the first year of service in Table C.4, we see that there are 84,494 accessions into the enlisted force annually—in the steady-state world, each year is like its predecessor and its successor, hence 84,494 accessions annually. We also see that in YOS 2 there are only 27,871 E1–3s remaining, and also that there are 34,669 E4s. The 34,669 E4s in YOS 2 must have come from the 84,494 E1–3s in YOS 1, and Table C.5 shows this same number of promotions into E4/YOS 2.

Table C.6

Aggregate Losses from Grade and YOS

YOS	E1–3	E4	E5	E6	E7	E8	E9	Total
1	21,954							21,954
2	4,094	2,198						6,293
3	311	17,961	2,265					20,537
4	310	4,079	7,922	1				12,312
5	36	1,716	2,741	39				4,532
6	7	1,409	1,933	182				3,530
7		847	1,336	396				2,579
8		417	950	619				1,986
9		53	430	679	4			1,166
10		34	182	654	53			923
11			7	611	89			707
12			6	410	87			504
13			6	280	128			414
14			5	159	115			279
15			26	100	88			213
16				56	67	8		131
17				38	59	6		104
18				22	52	23		97
19				1	25	4		30
20				5	199	46	1	251
21					3,274	1,043	23	4,341
22					7	325	48	380
23						238	66	304
24						472	56	528
25							72	72
26							53	53
27							68	68
28							46	46
29							39	39
30							123	123
Total	26,712	28,714	17,808	4,252	4,246	2,166	596	84,494

The remainder of the E1–3/YOS 1 enlisteds,

$$21{,}954 = 84{,}494 - 27{,}871 - 34{,}669,$$

must have left the Army, and Table C.6 confirms this.

The conservation-of-flow principle applies to any grade/YOS cell, and it can be used to verify that the model is operating properly. The model's implementation tests this principle for each grade/YOS cell, as well as other mathematical identities that must apply. A convenient panel provides the model user with warnings when these identities are violated.

CMF 67 (AIRCRAFT MAINTENANCE) IPM MODEL

CMF 67 base case model inputs and outputs appear in Tables C.7 through C.12. The inputs also derive from MOSLS run M9712.

The retention improvement and RCP relaxation alternatives derive from adjustments to the base case loss rates—see the overlay panels in Table C.8. For example, the 25 percent mid-career retention improvement case has E5 and E6 loss rates in years 7–10 set to 0.75 of the base case loss rates. The 50 percent improvement case multiplies the base case loss rates by 0.5. The 50 percent RCP relaxation case modifies E7 and E8 loss rates by setting E7 year 21 and E8 year 23 loss rates to 50 percent, followed by two 10 percent years and ending with 100 percent. The 90 percent relaxation case uses 10 percent instead of 50 percent.

Table C.7

CMF 67 (Aircraft Maintenance) Inventory by Grade

Total	E1–3	E4	E5	E6	E7	E8	E9
15,262	3,377	5,016	2,927	2,187	1,271	418	66

Table C.8

CMF 67 (Aircraft Maintenance) Loss Rates Out of Grade (percent)

YOS	E1–3	E4	E5	E6	E7	E8	E9
1	17.3						
2	10.2	5.4					
3	34.8	7.2	64.8				
4	79.0	11.0	15.0				
5	83.3	11.4	6.8				
6	100.0	48.2	49.4	38.1			
7		18.5	13.6	11.6			
8		31.4	20.1	13.2			
9		68.8	18.2	11.3			
10		100.0	18.6	14.5	2.4		
11			15.9	10.7	11.1		
12			11.7	8.5	8.0		
13			16.9	5.0	5.9		
14			17.3	3.8	5.2		
15			100.0	2.3	1.6		
16				4.5	3.0		
17				2.0	1.2		
18				1.3	0.6	2.1	
19				0.9			
20				100.0	2.0	1.5	
21					95.0	44.2	11.1
22					100.0	29.2	4.5
23						33.7	12.0
24						100.0	6.7
25							20.0
26							19.0
27							25.0
28							50.0
29							
30							100.0

25% retention improvement

YOS	E5	E6
7	10.2%	8.7%
8	15.1%	9.9%
9	13.7%	8.4%
10	14.0%	10.8%

50% retention improvement

YOS	E5	E6
7	6.8%	5.8%
8	10.1%	6.6%
9	9.1%	5.6%
10	9.3%	7.2%

50% RCP relaxation

YOS	E7	E8
21	**50%**	
22	10%	
23	10%	
24	100%	**50%**
25		10%
26		10%
27		100%

The base case loss rates are shown in the table, and the loss rates associated with mid-career retention improvement and RCP relaxation are shown in the overlay panels. The 90 percent RCP relaxation case is not shown, it differing from the 50 percent panel with the replacement of the two **50 percent** entries with 90 percent.

Table C.9a

CMF 67 (Aircraft Maintenance) Unadjusted Promotion Distributions Into Grade and YOS (percent)

YOS	E1–3	E4	E5	E6	E7	E8	E9
1	100.0						
2		6.8					
3		67.8	4.8				
4		23.4	10.1				
5		1.5	32.3	3.2			
6		0.4	26.1	7.6			
7		0.1	15.2	13.7			
8			5.7	12.0			
9			3.6	14.3	0.8		
10			1.9	15.9	2.9		
11			0.2	10.4	5.7		
12				8.1	8.2		
13				5.0	8.5		
14				3.4	10.2		
15				3.7	11.2	6.0	
16				1.5	11.7	5.7	
17				1.1	11.4	12.8	
18				0.1	10.0	16.4	3.5
19					8.5	17.4	7.0
20					7.4	17.4	14.0
21					3.5	13.4	15.8
22						8.1	22.8
23						2.7	15.8
24							12.3
25							8.8
26							
27							
28							
29							
30							
	100	100	100	100	100	100	100

Table C.9b

CMF 67 (Aircraft Maintenance) Adjusted Promotion Distributions Into Grade and YOS (percent)

YOS	E1–3	E4	E5	E6	E7	E8	E9
1	100.0						
2		67.0					
3		33.0	18.5				
4			39.2				
5			32.3	16.7			
6			10.0	39.3			
7				13.7			
8				12.0			
9				14.3	1.7		
10				4.0	6.3		
11					12.2		
12					17.5		
13					18.2		
14					10.2		
15					11.2	12.0	
16					11.7	11.3	
17					11.0	25.3	
18						16.4	5.3
19						17.4	10.7
20						17.4	21.4
21							24.0
22							22.8
23							15.8
24							
25							
26							
27							
28							
29							
30							
	100	100	100	100	100	100	100

Table C.10

CMF 67 (Aircraft Maintenance) Inventory by Grade and YOS

YOS	E1–3	E4	E5	E6	E7	E8	E9	Total
1	2,480							2,480
2	775	1,276						2,051
3	68	1,602	234					1,904
4	44	990	579					1,613
5	9	472	813	88				1,382
6	2	291	678	294				1,266
7		151	271	254				677
8		123	171	288				582
9		84	62	321	4			471
10		26	29	292	17			365
11			24	223	44			291
12			20	161	77			258
13			18	107	111			236
14			15	80	127			221
15			12	52	135	10		209
16				25	149	19		194
17				0	148	40		188
18				0	133	53	1	186
19				0	118	65	2	184
20				0	103	77	4	184
21					101	73	7	181
22					5	38	8	52
23						26	10	35
24						17	8	25
25							8	8
26							6	6
27							5	5
28							4	4
29							2	2
30							2	2
Total	3,377	5,016	2,927	2,187	1,271	418	66	15,262

Table C.11

CMF 67 (Aircraft Maintenance) Accessions and Promotions Into YOS

YOS	E1–3	E4	E5	E6	E7	E8	E9	Total
1	2,480							2,480
2		1,276						1,276
3		629	234					863
4			497					497
5			409	88				497
6			127	207				333
7				72				72
8				63				63
9				75	4			79
10				21	14			35
11					27			27
12					38			38
13					40			40
14					22			22
15					25	10		34
16					26	9		35
17					24	21		45
18						14	1	14
19						14	1	15
20						14	2	17
21							3	3
22							2	2
23							2	2
24								
25								
26								
27								
28								
29								
30								
Total	2,480	1,905	1,267	526	219	82	11	6,489

Table C.12

CMF 67 (Aircraft Maintenance) Separations Out of YOS

YOS	E1–3	E4	E5	E6	E7	E8	E9	Total
1	429							429
2	79	68						147
3	24	115	152					291
4	35	109	87					231
5	8	54	55					117
6	2	140	335	112				589
7		28	37	30				95
8		39	34	38				111
9		58	11	36				105
10		26	5	42	0			74
11			4	24	5			33
12			2	14	6			22
13			3	5	7			15
14			3	3	7			12
15			12	1	2			15
16				1	4			6
17				0	2			2
18				0	1	1		2
19				0				0
20				0	2	1		3
21					96	32	1	129
22					5	11	0	17
23						9	1	10
24						17	1	18
25							2	2
26							1	1
27							1	1
28							2	2
29								
30							2	2
Total	575	638	741	307	137	71	11	2,480

REFERENCES

Winkler, John D., et al., *Future Leader Development of Army Noncommissioned Officers: Workshop Results,* Santa Monica, CA: RAND, CF-138-A, 1998.

Management Sciences Group of GRC International's Military Personnel Operations Division, *Documentation Updates of Mathematically Complex Programs in ELIM, MOSLS, and OPALS, Documentation Work Products Volumes 1 and 2, GRC International, Inc.,* Vienna, VA, September 1996. (These volumes were prepared for the U.S. Army DCSPER's Military Strength Analysis and Forecasting Directorate as part of the effort to document the mathematical basis for the inventory projection models used by the directorate.)